Learning ArcGIS Runtime SDK for .NET

Learn how to build native, cross-platform mapping apps with this comprehensive and practical guide, using the MVVM pattern

Ron Vincent

[PACKT] open source*

PUBLISHING   community experience distilled

BIRMINGHAM - MUMBAI

Learning ArcGIS Runtime SDK for .NET

First published: June 2016

Production reference: 1240616

Published by Packt Publishing Ltd.
Livery Place
35 Livery Street
Birmingham B3 2PB, UK.

ISBN 978-1-78588-545-7

www.packtpub.com

Credits

Author
Ron Vincent

Reviewer
Prasad Lingam
Shaik Shavali

Commissioning Editor
Veena Pagare

Acquisition Editor
Manish Nainani

Content Development Editor
Aishwarya Pandere

Technical Editor
Jayesh Sonawane

Copy Editor
Safis Editing

Project Coordinator
Nidhi Joshi

Proofreader
Safis Editing

Indexer
Monica Ajmera Mehta

Graphics
Kirk D'Penha
Jason Monteiro

Production Coordinator
Melwyn Dsa

Cover Work
Melwyn Dsa

About the Author

Ron Vincent graduated with a B.S. in Geography from the University of Utah in 1992. Since then, he has worked in applying GIS to many organizations across many industries for 24 years. His primary experience involves building GIS solutions for the U.S. military, which includes the U.S. Army, Navy, and Marine Corps. It also includes building custom solutions for utilities, mining, healthcare, banking, commercial business, national parks, and so on. This has amounted to 100+ projects related to implementing Esri's GIS technologies. Because of a variety of industries and projects, Mr. Vincent has had the opportunity to be a developer, analyst, consultant, trainer, and manager of the GIS projects. Over the years, he has developed and conducted both formal and informal training sessions. This includes training parcel mappers in the local government, GIS analysts in many industries, soldiers in the U.S. and foreign militaries, and junior and senior developers using a wide range of technologies.

Mr. Vincent lives in North Carolina with his wife Sandy. He has four children (Naudia, Spenser, Zach, and Kjersten). While not working, he enjoys reading (software books, of course) and traveling around this awesome world we live on. For more information, please see his LinkedIn page: `https://www.linkedin.com/in/ron-vincent-8958145`.

About the Reviewers

Prasad Lingam has been passionately exploring Geoinformation technologies for almost 10 years. He has gained knowledge in the application of Geoinformatics to areas such as urban planning, transportation, utilities, environment and construction management, thus leveraging his civil engineering background. He is currently working in the water and waste water domain, implementing geospatial analysis in Desktop GIS and promoting web as well as mobile-based GIS applications for solving operational and planning issues. His work experience spans to project locations such as Perth, the Middle East, New Zealand, Fiji, and India. He is keen toward studying the confluence of geospatial technologies with technologies such as Big Data, BIM, Geo-Visualization, and so on.

Shaik Shavali, a senior GIS developer at Dar Al-Riyadh, Shaik, has 7 to 8 years of experience in the field of geospatial technologies and projects. His areas of expertise are developing custom GIS web and mobile applications using the latest ESRI technologies. He was one of the lead developers for Emergency Response Management Systems for the largest Islamic pilgrimage (Hajj), gathering nearly 2 million people. Currently, he is actively taking part in designing and developing GIS projects for the government sector in Saudi Arabia.

He has received his bachelors in engineering. Presently, he is pursuing his masters in GIS through UNIGIS.

Firstly, I would like to thank Allah for his countless blessings. I would also like to thank my parents (Akbar Saheb and Fathima), in-laws (Ehasanulla and Shahnaz), and family for their emotional support, guidance, and prayers. Finally, I would like to thank my better half, Farheen Ehasanulla, for her love and support, which has always pushed me to do better. She is my backbone, best critic, and most importantly, my best friend.

www.PacktPub.com

eBooks, discount offers, and more

Did you know that Packt offers eBook versions of every book published, with PDF and ePub files available? You can upgrade to the eBook version at www.PacktPub.com and as a print book customer, you are entitled to a discount on the eBook copy. Get in touch with us at customercare@packtpub.com for more details.

At www.PacktPub.com, you can also read a collection of free technical articles, sign up for a range of free newsletters and receive exclusive discounts and offers on Packt books and eBooks.

https://www2.packtpub.com/books/subscription/packtlib

Do you need instant solutions to your IT questions? PacktLib is Packt's online digital book library. Here, you can search, access, and read Packt's entire library of books.

Why subscribe?

- Fully searchable across every book published by Packt
- Copy and paste, print, and bookmark content
- On demand and accessible via a web browser

Table of Contents

Preface

Welcome to the exciting world of building native, cross-platform mapping apps! For many kinds of app, a map is not only a nice feature, it's a must. With this book, you're going to explore ArcGIS Runtime, which is a Software Developer Kit (SDK) from Esri, the world's largest provider of Geographic Information System software. With this SDK, you're going to be taken on a journey from the basics up to an intermediate level of understanding how to build a complete, professional application by learning the ArcGIS Runtime SDK and Application Programming Interface (API). With ArcGIS Runtime, you can build native, cross-platform mapping apps that can include the ability to display data in 2D and 3D from online or offline data sources, edit online and offline data, online and offline geocoding and routing, conduct spatial analysis, navigation, and so on. What's more is that ArcGIS Runtime comes in several languages, such as .NET (C# and Visual Basic), Java SE, Objective-C, and C++. As a result, you can build highly performant apps and run them natively on Windows, iOS, OS X, Android, and Linux.

With the API, you can build interactive, beautiful maps that you can then sell on the App Store, or make available to users in your organization as part of an Enterprise Mobility strategy, or even embed a map into another app such as a military system. Your users will then be able to access a rich set of content from the cloud (ArcGIS Online or Portal for ArcGIS) or in disconnected mode using offline data that you or your organization has created. As you gain more familiarity with the API, you'll find that it's much easier than previous technologies from Esri and offers new possibilities such as augmented reality, real-time mapping, virtual reality, disconnected analytics, and a whole host of other possibilities.

Although Esri's help for ArcGIS Runtime is great, you'll find that this book offers more. In fact, this book will provide you step-by-step instructions in C# for building an app. Not only that, it shows you how to structure your app using the Model-View-ViewModel pattern, a popular pattern for .NET developers. You will, along the way, learn about 3D mapping and spatial analysis and even how to test your apps using test-driven development and take a deep dive into performance testing. By the time you're done reading this book, you will have all the necessary skills to build an app from beginning to deployment on the selected devices, no matter if it's a laptop, tablet, or smart phone. The reason for this is that the API is nearly identical across all platforms. As a result, what you learn in this book, except for differences in the languages, will for the most part apply to other flavors of ArcGIS Runtime. Learn the API once, and then apply across multiple platforms.

What this book covers

Chapter 1, Introduction to ArcGIS Runtime, introduces you to the world of native mapping apps, where you will learn about the ArcGIS Runtime SDK language options, the history of Esri's developer technologies, native versus web versus hybrid apps, types of mappings apps, ArcGIS Runtime architecture, system requirements, setting up your development environment, and finally, you'll build your first app using a code-behind file approach.

Chapter 2, The MVVM Pattern, takes the app you built in *Chapter 1, Introduction to ArcGIS Runtime*, and transforms it using a pattern called Model-View-ViewModel. With this pattern, you'll have an app that supports the notion of separation of concern, which makes your app more maintainable, blendable, and testable. The concepts you will learn in this chapter will be expanded on as you walk through the remaining chapters. At the same time, you will explore the ArcGIS Runtime SDK.

Chapter 3, Maps and Layers, introduces you to the MapView class and how to add all the different kinds of layers supported by ArcGIS Runtime. You will learn about projections, MapGrid, scale, overlays, location display for navigation, events, map interactions, map extent, the wide range of supported layers types (dynamic map service, tiled map service, feature services, and so on), the ArcGIS Runtime geodatabase, and so on.

Chapter 4, From 2D to 3D, will introduce you the exciting world of 3D by explaining the importance of 3D, how to navigate in 3D, how to add layers in 3D, adding vertical exaggeration using surface models, MVVM in 3D, and 3D content.

Chapter 5, Geometry and Symbology, will take you down to the fundamental constructs of layers by explaining all of the options you have to create geometry, and then how to symbolize the geometry in your layers. You will learn the difference between mutable and immutable geometry for the MapPoint, envelope, multipart, polyline, polygon, and multipoint geometry, and then explore the GeometryEngine class so that you can do vector operations, such as reprojecting and union. You'll also learn about converting units of measure and area. After geometry is covered, you will then learn about how to symbolize the geometry using simple symbols, renderers, and even military symbols.

Chapter 6, Displaying Information, teaches you how to handle mouse events with MVVM so that you can display information by interacting with the map. You will also learn how to make your maps even more useful by showing a legend, scale bar, label features, create overlays, and showing your current location using a GPS.

Chapter 7, Finding, Querying, and Identifying Features, will introduce you to tasks so that you can find objects in the map, conduct online and offline searching, and interactively discover information in a layer by just clicking on an object (identify).

Chapter 8, Geocoding and Routing, provides you with an understanding of geocoding, which is the process of turning a street address into coordinates. You will learn about online and offline geocoding, finding places using just their name, and then how to route between two or more locations.

Chapter 9, Editing Features, details the differences between online and offline editing, how to prepare data for either online or offline editing, and then how to add, update, and delete features. This chapter also discusses the concept of selection, and finally, committing and syncing edits.

Chapter 10, Spatial Analysis, will introduce you the spatial analysis using ArcGIS Runtime, and then discuss geoprocessing, which allows you to exploit the powerful tools from ArcGIS Desktop or Pro so that you can add powerful analytical capabilities to your app from either ArcGIS Server or locally using a geoprocessing package. You learn, by way of example, how to do line of sight analysis and generate drive-time polygons. Also, you'll learn about synchronous and asynchronous geoprocessing.

Chapter 11, Testing and Performance, will show you how to take the concepts you learned about in the previous chapters with regards to the API and MVVM by showing you how to unit test your app so that you know it works at the most fundamental level. You will also be shown the important of test-driven development using a sample app you developed in an earlier chapter. Lastly, you will take a deeper dive into optimizing your app from the server side if it uses on-premise data sources. You will then get a detailed explanation of how ArcGIS Runtime's high-performance rendering engine works.

Chapter 12, Configuring, Licensing, and Deploying, will finally explain what you will need to consider when configuring your app so that users can make the app look and behave how they need it to, and then discuss what steps you need to take to license your app properly, and finally, the steps required to deploy your app to your users.

What you need for this book

To work through the sample apps in this book, you will need a reasonably good laptop or desktop computer that meets these minimum hardware/OS/.NET requirements:

- Internet access with a sufficiently fast download speed.
- 2.2 GHz or higher.
- Intel Pentium, Intel Core Duo, or Xeon Processor.
- 1.5 GB of available disk space.
- 2 GB or higher of RAM. Even more may be required for the chapter on 3D.
- A graphics card with 64 MB of RAM, 256 MB of RAM is recommend. NVIDIA, ATI, and Intel chipsets are supported.
- Windows 8.1 Basic, Professional and Enterprise, Windows 7 Ultimate, Enterprise, Professional or Home Premium, Windows 7 SP1 Ultimate, Enterprise, and Professional or Home Edition.
- .NET 4.5.2, 4.5.1 or 4.5. The sample apps are using .NET 4.5.
- Visual Studio 2013 (all editions). You can also use Visual Studio Express 2013 for Windows Desktop, Visual Studio 2012 with Update 3 (all editions), and Visual Studio Express 2012 for Windows Desktop and Update 3. *Chapter 1, Introduction to ArcGIS Runtime*, will show you how to download a free copy of Visual Studio 2013. It's also possible to use Visual Studio 2015, but ArcGIS Runtime does not currently have a project template for that version.

Who this book is for

This book is for you. It will take your through a journey of how to use the API and provide you with enough details so that you can apply these concepts to the other supported languages. However, this book will focus on the .NET version of the SDK because it offers all of the capabilities due to the fact that it runs on a Windows desktop or laptop computer. This book will focus on C# because it is quite popular with the .NET developers and because many long term Esri users have a great deal of experience with previous technologies, such as ArcGIS Desktop and Engine.

This doesn't mean you will only understand the material if you're familiar with Esri's technology. In fact, just the opposite is true. If you're new to Esri technology, you will be introduced to the concepts and then shown where to go to learn more about Esri's platform.

Most of the content was written for new to intermediate developers. However, there are a few cases where more advanced material is presented. This is especially the case if you're new to 3D in *Chapter 4*, *From 2D to 3D*, or unit testing and performance testing in *Chapter 11*, *Testing and Performance*. When it comes to .NET, you should have a basic understanding of C# and, in particular, of the task parallel library and asynchronous programming in general. You should also understand the basics of Windows Presentation Foundation, with a special emphasis on binding. All the other material will be explained so that you can build a complete native app.

Conventions

In this book, you will find a number of text styles that distinguish between different kinds of information. Here are some examples of these styles and an explanation of their meaning.

A block of code is set as follows:

```
// create a graphics layer and give it an ID
var graphicsLayer = new Esri.ArcGISRuntime.Layers.GraphicsLayer();
graphicsLayer.ID = "MyGraphicsLayer";

// add the graphics layer to the map
MyMapView.Map.Layers.Add(graphicsLayer);
```

A block of XAML code will appear like this:

```
<esri:GroupLayer DisplayName="Basemap Group">
    <esri:ArcGISTiledMapServiceLayer
        DisplayName="Imagery" IsVisible="False"
            ServiceUri="http://services.arcgisonline.com/ArcGIS/
                rest/services/World_Imagery/MapServer" />
    <esri:ArcGISTiledMapServiceLayer DisplayName="Street"
        ServiceUri="http://services.arcgisonline.com/ArcGIS/rest/
            services/World_Street_Map/MapServer" />
</esri:GroupLayer>
```

When we wish to draw your attention to a particular part of a code block, the relevant lines or items are set in bold:

```
USE MyDatabase;
GO
SELECT Name, ProductNumber, ListPrice AS Price
FROM Production.Product
WHERE ProductLine = 'R'
AND DaysToManufacture < 4
ORDER BY Name ASC;
GO
```

New terms and **important words** are shown in bold. Words that you see on the screen, for example, in menus or dialog boxes, appear in the text like this: "Clicking the **Next** button moves you to the next screen."

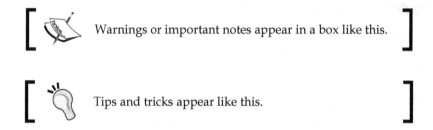

Warnings or important notes appear in a box like this.

Tips and tricks appear like this.

Reader feedback

Feedback from our readers is always welcome. Let us know what you think about this book—what you liked or disliked. Reader feedback is important for us as it helps us develop titles that you will really get the most out of.

To send us general feedback, simply e-mail feedback@packtpub.com, and mention the book's title in the subject of your message.

If there is a topic that you have expertise in and you are interested in either writing or contributing to a book, see our author guide at www.packtpub.com/authors.

Customer support

Now that you are the proud owner of a Packt book, we have a number of things to help you to get the most from your purchase.

Downloading the example code

You can download the example code files for this book from `https://github.com/rvinc66/ArcGISRuntimeBook`.

You can download the example code files for this book from your account at `http://www.packtpub.com`. If you purchased this book elsewhere, you can visit `http://www.packtpub.com/support` and register to have the files e-mailed directly to you.

You can download the code files by following these steps:

1. Log in or register to our website using your e-mail address and password.
2. Hover the mouse pointer on the **SUPPORT** tab at the top.
3. Click on **Code Downloads & Errata**.
4. Enter the name of the book in the **Search** box.
5. Select the book for which you're looking to download the code files.
6. Choose from the drop-down menu where you purchased this book from.
7. Click on **Code Download**.

You can also download the code files by clicking on the **Code Files** button on the book's webpage at the Packt Publishing website. This page can be accessed by entering the book's name in the **Search** box. Please note that you need to be logged in to your Packt account.

Once the file is downloaded, please make sure that you unzip or extract the folder using the latest version of:

- WinRAR / 7-Zip for Windows
- Zipeg / iZip / UnRarX for Mac
- 7-Zip / PeaZip for Linux

Downloading the color images of this book

We also provide you with a PDF file that has color images of the screenshots/diagrams used in this book. The color images will help you better understand the changes in the output. You can download this file from `http://www.packtpub.com/sites/default/files/downloads/LearningArcGISRuntimeSDKForNET_ColorImages.pdf`.

Errata

Although we have taken every care to ensure the accuracy of our content, mistakes do happen. If you find a mistake in one of our books—maybe a mistake in the text or the code—we would be grateful if you could report this to us. By doing so, you can save other readers from frustration and help us improve subsequent versions of this book. If you find any errata, please report them by visiting http://www.packtpub.com/submit-errata, selecting your book, clicking on the **Errata Submission Form** link, and entering the details of your errata. Once your errata are verified, your submission will be accepted and the errata will be uploaded to our website or added to any list of existing errata under the Errata section of that title.

To view the previously submitted errata, go to https://www.packtpub.com/books/content/support and enter the name of the book in the search field. The required information will appear under the **Errata** section.

Piracy

Piracy of copyrighted material on the Internet is an ongoing problem across all media. At Packt, we take the protection of our copyright and licenses very seriously. If you come across any illegal copies of our works in any form on the Internet, please provide us with the location address or website name immediately so that we can pursue a remedy.

Please contact us at copyright@packtpub.com with a link to the suspected pirated material.

We appreciate your help in protecting our authors and our ability to bring you valuable content.

Questions

If you have a problem with any aspect of this book, you can contact us at questions@packtpub.com, and we will do our best to address the problem.

1
Introduction to ArcGIS Runtime

In this chapter, you will be introduced to ArcGIS Runtime. ArcGIS Runtime is a powerful mapping **Software Developer Kit** (**SDK**) that allows you to create native apps on many platforms, such as Windows, iOS, and Android, using a variety of languages, such as .NET, Java SE, Objective-C, and C++ with Qt. The purpose of the chapter is to take you on a tour of ArcGIS Runtime and show you how to build an app using the ArcGIS Runtime SDK for .NET.

In summary, the following topics will be discussed:

- Providing high-level examples of the use of ArcGIS Runtime
- Language choices
- A little history of the development tools of Esri (www.esri.com)
- Native apps across the broader market
- Types of mapping apps
- The architecture of ArcGIS Runtime
- Areas where ArcGIS Runtime fits within the Esri platform
- Installation steps of the ArcGIS Runtime SDK for .NET
- Building your first app with ArcGIS Runtime
- Configuring, testing, and structuring your app

Introducing ArcGIS Runtime

Get ready to build awesome native apps! In this chapter, you will be introduced to ArcGIS Runtime in general. You will be presented with an overview of where it fits within the Esri platform, whom it is for, what problems it helps you solve, and how to get started; you will be presented with an overview of where it fits within Esri's platform, whom it is for, what problems it helps you solve, how to get started, not only with Runtime, but also the other tools you'll need to be successful at building native mapping apps.

What is ArcGIS Runtime? Well, to put it simply, it is a powerful mapping SDK for building native apps. However, it goes beyond just simple mapping because it is a core developer technology from Esri, the world's largest software company for **Geographic Information Systems (GIS)**. Basically, you can easily make a map, add layers, and do analysis, such as buffer a point or line, visibility analysis, online and offline editing, and find the quickest route between two locations, which is really what separates Esri's technology from other mapping technologies. ArcGIS Runtime can be used to create the native mapping apps on iOS, OS X, Android, Linux, and Windows. It was developed in C++ and it has been designed to be as fast and efficient as possible on each OS it runs on. It's cross-platform, 64-bit, multithreaded, has a much simpler API than previous technologies from Esri, and it works with the local and cloud data sources. More importantly, for you, the developer is such that you don't have to make a language choice, and then go out and learn it before you can start developing. Esri has made this part easy for you as the developer so that you can use whatever language you're comfortable with and start developing immediately. However, if you're wondering whether you will make a sacrifice if you choose one OS over another, don't worry. The **Application Programming Interface (API)** is nearly identical across all supported operating systems. In short, you get the functionality you need no matter the language or platform you choose. What more could the developer ask for?

ArcGIS Runtime can be used in a variety of ways. These are the following three broad categories where ArcGIS Runtime can be used:

- App stores
- Embedded systems
- Enterprise mobility

ArcGIS Runtime can be used for developing apps that you sell on Google Play, Apple App Store, or Windows Store. For example, it could provide the citizens of a city, town, or even nation with a phonebook that shows not only the map but also the directions from your current location to your favorite restaurant even if you don't have Internet access. See this screenshot:

ArcGIS Runtime can also be used to embed into another app. A perfect example of this is to embed it into a military vehicle real-time tracking system. This has been done with older technologies from Esri called ArcGIS Engine (Refer to *A little history* later in this chapter). See this screenshot:

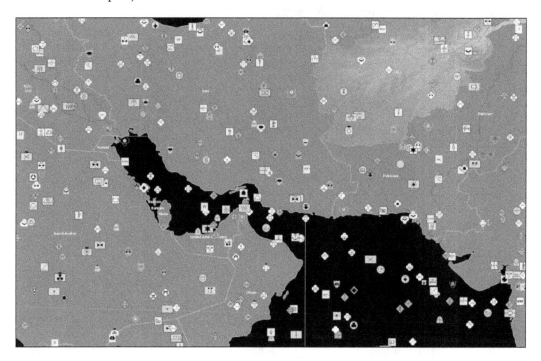

Another use of ArcGIS Runtime could be a mobile solution for a business' enterprise effort to go mobile (enterprise mobility). For example, a utility may need to go out and perform damage assessment after a storm. As it's very possible that cell towers may be down, there is a real need to be able to find a location, add some data to the map (where the fallen debris is located) so that field crews can estimate how much work needs to be done.

What can ArcGIS Runtime do? Here's a quick summary:

- Online and offline geocoding
- Online and offline routing
- Online and offline search (both tabular and spatial queries)
- Online and offline analysis
- Online and offline editing

Well, it can certainly make a map using both online and offline content. With these maps, you will be able to find things such as streets, points of interest, people, and equipment, using a variety of methods to find objects on the map either by just clicking on the map, entering coordinates, or an address. You'll be able to then route yourself to the desired location(s). However, it goes beyond simple queries that you'd find in any database system. With ArcGIS Runtime, you can also do spatial searches such as finding all of the manholes within 1 mile of this location using a simple one-mile circle (buffer). What if you're developing an app for a business' fleet of vehicles where the drivers not only want to visit multiple sites during the day, but also want to find other potential customers along the route so that they can maximize their time and minimize the fuel consumed? Well, that's just one of the many spatial analytical operations that can be built into the app. An even more critical app example is a soldier that needs to go into hostile territory where GPS is denied, but they need to make observations by adding enemy locations to a map while completely disconnected, and then find the quickest route back to base across rough terrain. Another military example is a battlefield commander who not only needs to have situation awareness by tracking friendly forces in real time while they're at the **Forward Operating Base (FOB)**, but they also need to see it while they step outside or drive to another location.

For those who are more experienced Esri users, here's a more detailed summary of its features:

- High performance in 2D and 3D
- Drawing, querying, identifying, and selecting features and graphics
- Using Portal for ArcGIS, ArcGIS Server, and ArcGIS Online
- Working offline with local basemaps and data
- TPKs, mobile GDB, shapefiles, KML, and CSV
- Editing and synchronizing geocode and routing
- Working with device sensors
- Performing advanced geometric operations locally
- Task-based asynchronous pattern
- Model-View-ViewModel friendly

Language options

What are the language options? With ArcGIS Runtime, you can choose either .NET, Java SE, Android, Objective-C, C++ using Qt, or a combination of C++ and JavaScript by using Qt and QML. The beauty of this feature is that it's completely up to you and your team of developers. If you work alone, you probably already have a language in mind, and based on the title of this book, you're probably a .NET developer. If you're not a .NET developer, the good news is that the things you learn here will apply to the other flavors of ArcGIS Runtime. There will be some minor differences in how Runtime is implemented in your language of choice, however. With that said, which language choices do you have based on the OS? Refer to the following diagram. On the desktop, you can use Objective-C, .NET, C++ via Qt and QML, and Java SE. On mobile devices, you can use Objective-C, Java with the Android SDK, or .NET for Windows Phone. For embedded apps, you can use Java SE, C++ via Qt and QML, or .NET:

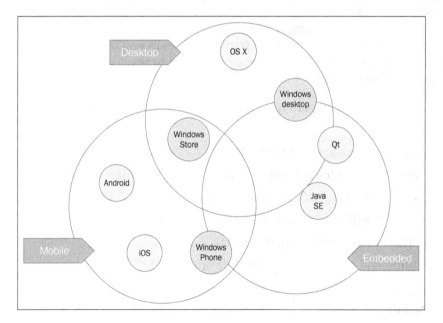

One of the awesome benefits of many of these languages is that you can build apps that are truly cross-platform. From the software development perspective, this means you can write your code once and run it on many platforms. For example, it's possible to use the .NET SDK to write an app that shares code and can be run on Windows, Windows Store, and Windows Phone. With the Qt version of ArcGIS Runtime, you can go even further and write your code so that it runs on nearly all platforms. That's amazing! But don't feel left out. Later this year, Esri will be releasing the Xamarin® version of ArcGIS Runtime, which will allow you to run your .NET apps on iOS and Android too. As a result, you get more bang for your buck.

A little history

Esri has a long history of providing powerful capabilities to its users for server, desktop, and mobile platforms. It really started back in the 1980s with ArcInfo and with a language called **Arc Macro Language** (**AML**). This language allowed users to create scripts that automated tasks and provided developers with the ability to create interfaces on both Windows and UNIX. Then, PCs became more powerful and Esri continued providing great developer opportunities with AML and new programs that ran on Windows called **ArcView** and **MapObjects**. Avenue was the first attempt of Esri at an object-oriented programming language. It was very popular because it provided its user base with their first exposure to this new paradigm of writing code. Of course, by the late 1990s, everything started to change with languages such as Visual Basic and **Visual Basic for Applications** (**VBA**). Not only that, languages such as Java became very popular, which led to .NET. As Esri saw these major developments, with more high-level languages that were easier for its user base, it became more and more obvious that developing a custom language for their user base no longer made sense. By the late 1990s, Esri started building ArcGIS, which was going to allow users to use their language of choice (C++, .NET, Java, VB, and VBA). Users could then develop a tool or extension for the flagship product of Esri (ArcGIS). Esri also decided to provide users with even more customization options with a very powerful product called ArcGIS Engine, which gave developers the full capability to customize every aspect of their flagship product. Since around the year 2000 and up until about 2014, ArcGIS Desktop and ArcGIS Engine were the main systems that most developers used to customize Esri's technology. The API was called **ArcObjects** and it was written in C++ and Microsoft's **Component Object Model** (**COM**). All these technologies evolved as the computer industry evolved and Esri, to its credit, kept pace despite the complexity and size of their GIS.

When it comes to web apps, Esri saw its first opportunity to leverage MapObjects, which was for Visual Basic, to allow users to take their work and push it to the web. As a result, they created **MapObjects Internet Map Server** (**MOIMS**). I actually had a great opportunity to create the first local government MOIMS app for Cabarrus County, NC back in the late 1990s. Back then, web mapping was pretty bare bones, but it worked! From there, ArcIMS came along, and then ArcGIS for Server was introduced. To aid web developers, Esri built their JavaScript API, which really brought about an incredible explosion of web apps over the past several years. All of this eventually led users to need a common platform to present their web content and apps in a portal. Eventually, **ArcGIS Online** (**AGOL**) and the on-premise version of AGOL, called Portal for ArcGIS, came about, which allowed users to create maps and apps very quickly. More importantly, this brought GIS to the masses.

Another important development over the past few years with Esri was the development of mobile solutions such as ArcPad and ArcGIS for Windows Mobile. Despite the success of these products, it is clear that ArcGIS Runtime is the future of mobile apps for the technology stack of Esri because of the rise of so many platforms, such as Android, Apple, and Microsoft-supported devices. As such, Esri is developing several client-side apps with ArcGIS Runtime for end users, such as Operations Dashboard for ArcGIS, Collector for ArcGIS, Navigator for ArcGIS, and ArcGIS Earth.

Apps

Despite the explosion of web apps, there still remains a need for native apps because if the Internet isn't available, many application scenarios simply aren't satisfied unless the user has the ability to work offline. As such, native apps are in abundance and growing. According to the site `https://www.statista.com/chart/3530/app-store-growth/`, native iOS apps are growing at a rate of 1,000 apps per day, as shown here:

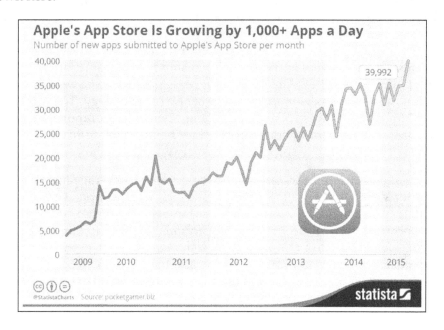

That's phenomenal! Even more interesting is that people use native apps more than web apps, according to `businessinsider.com`, as shown here:

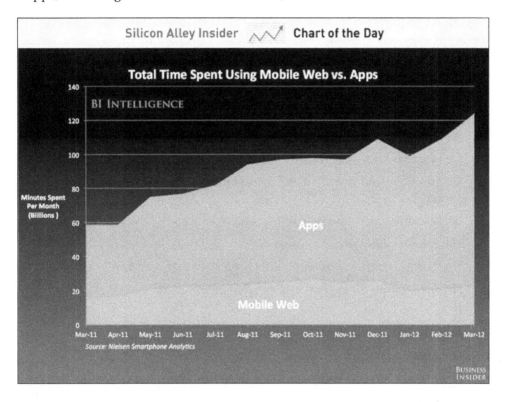

However, here's a question: Should a developer build a native app or a web app? Which is most appropriate? On the one hand, everyone wants their app to be functional, useable, reliable, performant, and scalable, right? Besides, do you want to create a web app, native app, or hybrid app? It's pretty common that building a native app takes longer and costs more than building a web app. Yet, web apps aren't as responsive and still don't work, and will never work if you don't have a connection to the Web. So, which approach should you take? (Note that it's possible to cache web content on a device, but that has its limitations.)

Native versus web versus hybrid

Deciding on whether to develop a web or native app is the most important decision you'll make once the app has been approved to be developed by your users or customer. In some cases, this is something already decided by the user or customer because they already have something in mind. On the other hand, if you're building an app for public consumption, you and your team will have to make the decision. No matter what, it can sometimes be a difficult decision, so it's important you understand the factors. Here are some of the deciding factors:

- Are the users connected and/or disconnected?
- Do you need device sensors?
- Do you have online and/or offline data sources? Can you create offline content?
- What kind of analytical capabilities are required?
- What's your budget?
- Do your developers have the skills for a native or web app?
- How frequently does your app need to be updated?
- How large is the data?
- How many users are there?
- Do you need secure storage of data?
- How much time do you have to develop?
- How often does the data change?
- What kind of expertise do you need?
- Can your users easily download the native app?
- Does the app need to do heavy processing? If so, do you have the processing power on the server, or if this is native, can the device handle it?
- Does your organization have a hardware platform policy?
- Will your users accept the cost?
- Is there a maintenance fee?

That's just a taste of the kinds of question that you need to carefully evaluate. Another important wrinkle in this decision is that sometimes you find you have to combine both web and native. As a developer, I have had to write an app for a large utility that needed to integrate into their enterprise asset system, which used a web app frontend but also required native performance due the volume of the data and the speed at which users drove down the road.

There are no easy answers here. You will need to carefully evaluate your alternatives and make the best decision. Don't be surprised if you start down one path and realize that the other path was more appropriate. If the answer isn't obvious, hire a consultant, review other apps (both native and web), determine the cost, get everyone involved, conduct a proof of concept, and most importantly, be prepared to learn and keep an open mind.

Types of mapping apps

Broadly speaking, there are the following four types of app that ArcGIS Runtime can be used for:

- Map-centric
- Non map-centric
- Nonmap with analytical tools
- Other

A map-centric app is an app that mostly shows a map with other windows and content focused around the map, such as the following screenshot. Note that the map takes up the entire window. All interface elements are on the map:

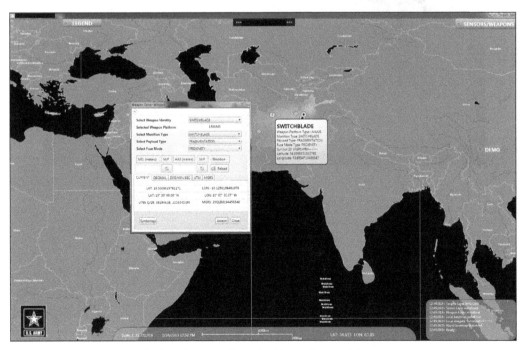

Display of a map with content focused around the map

A non map-centric app will contain a map, but it isn't the central focus of the app. The map simply supplements the primary purpose of the map. Refer to the University of Oregon app, where **Maps** is just one of many parts of a larger app. Finally, there are apps where you don't see a map at all, but it might use geocoding and/or routing in the background and just return a coordinate or directions in the form of a list:

Now, there are new types of map displays, such as the Pufferfish, which is a globe that renders a map on its spherical display, as shown here:

Map with a spherical display

Architecture of ArcGIS Runtime

As noted earlier, ArcGIS Runtime is written in C++ and compiled on each supported platform natively. Unlike ArcGIS Desktop and ArcGIS Engine, which relied heavily on COM, Runtime Core is pure C++. Therefore, it is as small and efficient as possible. The following architecture diagram shows the high-level architecture of ArcGIS Runtime. As you can see in the yellow box, there is **C++ Runtime Core**. This means that it makes full use of the hardware and graphics card of each OS. This aspect allows very fast performance for displaying the map content and analysis; it literally means that the API is pretty much the same across all platforms. In other words, Esri is maximizing code reuse, while at the same time, maximizing performance because it is natively compiled. Not only that, Runtime Core is actually pretty small. On Windows desktop, at the current release, it's only about 23 MB in size for the x86 build. For the x64 build, it's only about 30 MB. As such, it will fit easily on disk for any modern device. Esri has also developed this architecture so that it also minimizes battery use. Lastly, each API on top of Runtime Core has been designed to expose the full capabilities of Runtime Core.

When it comes to the rendering engine, Runtime Core uses OpenGL, OpenGL ARB shader, OpenGL ES 2.0, DirectX 9, and DirectX 11. Each rendering technology is selected to optimize the performance on each OS. Therefore, a desktop computer with a powerful graphics card can achieve maximum performance whereas a mobile device will obviously use a less powerful graphics card but it still maximizes the use of OpenGL ES, for example. Not shown in the architecture diagram is the **Hardware Abstraction Layer** (**HAL**) between the rendering engines and Runtime Core. With HAL, Esri can add even more rendering engine capabilities on top of these native rendering technologies, such as DirectX:

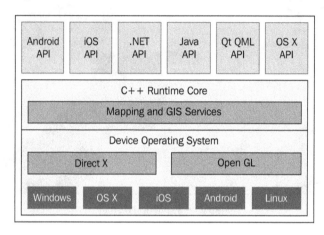

As a developer of ArcGIS Runtime, it's important to understand that the choices you make at the API level at the top of architecture really decide what happens at the HAL level and lower.

The Esri platform

Where does ArcGIS Runtime fit with the Esri platform and what do you need to really develop a native app? What is the meaning of the word *platform*? When an organization invests in Esri's technology, they are actually investing in something similar to a hardware platform. Think of it in this way: if your organization decides to go with the technology of Microsoft, they have deliberately chosen to purchase hardware that is compatible with the Windows OS, software that is compatible with Windows, the business applications of Microsoft, and so on. Well, this is effectively the same thing with the Esri platform. When an organization chooses the Esri platform, they are deciding to use the database formats of Esri, language choices, the very way Esri views the geospatial technology, and everything else that goes along with this choice.

ArcGIS Runtime is part of this platform. This doesn't mean you can use other technologies such as Google Earth, but it does mean that the organization has made a conscious decision to rely on a host of tools from Esri to satisfy their geospatial needs.

With that in mind, ArcGIS Runtime is a developer toolkit that you can use in and of itself, or as part of the Esri platform. If used within the context of the Esri platform, ArcGIS Runtime is a client app, as shown in the following diagram, in the top tier. However, you can also use ArcGIS Runtime without anything else from Esri. This means you don't need ArcGIS Desktop or Pro, ArcGIS Server, Portal for ArcGIS, AGOL Online, or any other technology. You can easily use publicly available web services or download shapefiles from publicly available sites and build an app. On the other hand, if you need to create your own data or build your analytical tools, you'll need one of these other technologies from Esri:

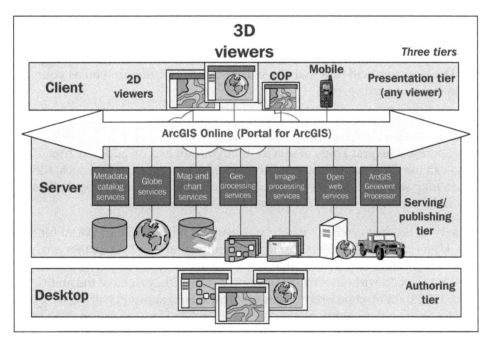

The preceding diagram will be found at `http://www.wiki.gis.com/wiki/index.php/File:SDS15Fig2.11_ArcGISSOA.jpg`.

ArcGIS Desktop (Desktop) is a very powerful desktop authoring technology that allows high-end GIS users to create data that includes complex editing, such as maintaining a complex electric utility network or editing a local government parcel database. It includes a wide variety of analytical tools, such as the ability to do routing, visibility analysis, line-of-sight analysis, spatial statistics, geostatistics, overlays, location-allocation analysis, and data management. In short, it allows you to create layers either from direct editing or as a result of analytical operations, and lastly, it allows you to publish your results (including analytical tools) to ArcGIS Server so that you can share it with others in your organization. You can even build scripts with Python or powerful tools with ArcObjects. Although ArcGIS Desktop has been around for about 15 years now, it is still the workhorse of Esri for the GIS power users. Despite this, it will eventually be replaced by its successors called ArcGIS Pro, which is the next natural step from the 32-bit ArcGIS Desktop to ArcGIS Pro, which is 64-bit technology.

ArcGIS Desktop and ArcGIS Pro are very important to the ArcGIS Runtime developer because they allow you to create layers and map objects that you may consume in your Runtime app. As ArcGIS Pro more and more begins to supplant ArcGIS Desktop, you will find that Pro will be more important to you as your requirements and needs grow.

After you've created layers and maps, you will need to make these applications available to ArcGIS Runtime. This is done with ArcGIS Desktop or Pro as a web service. There are several kinds of service that you can expose and that ArcGIS Runtime can use. You can publish and consume basemaps, map packages, tile packages, map servers, feature services, and so on. Throughout this book, these topics will be expanded on and consumed in the code examples.

Once a service has been created, it can be consumed with AGOL or Portal for ArcGIS (Portal). Users can add these services to their online maps and do further analysis with them. It should also be noted that users can also publish content on AGOL or Portal for ArcGIS without ArcGIS Desktop. On AGOL, you have the ability to create and use data in some pretty awesome ways. For example, with AGOL you can do Hot Spot analysis without ever needing ArcGIS Desktop or Pro. You can even add data to an online map in AGOL and ArcGIS Runtime can consume the resulting service. So, do you really need ArcGIS Desktop or Pro? Well, like anything, it depends. If you need sophisticated analytical or editing capabilities, Desktop or Pro is the technology to use. Otherwise, you may find that AGOL may be more than sufficient. In the end, each technology should be carefully evaluated along with the cost. If you're building a somewhat basic consumer product, you may find that AGOL will be more than sufficient. On the other hand, if you're part of an enterprise GIS that has invested in GIS technology and staff, it's makes perfect sense to use Portal, and Desktop, and Pro. Many organizations use Desktop/Pro, AGOL, Portal, custom web and desktop apps, and other third-party enterprise systems and databases, all in an effort to satisfy their users.

For more information about the Esri platform, refer to `http://www.esri.com/~/media/Files/Pdfs/products/arcgis-platform/architecting-the-arcgis-platform`.

SDK and API

Now that we've covered some background information about Esri's technology, let's return to ArcGIS Runtime. ArcGIS Runtime is the first and foremost SDK and an API. It's an SDK that provides you with the tools (Visual Studio templates, samples, and so on) so that you can quickly build apps. It exposes a rich API in the form of objects so that you can satisfy nearly any requirement your app needs. You can simply pick and choose what you need and decide how to implement the required solution. However, before you do that, you really need to know what you expect and the steps required to get started. As you go through the remaining parts of this book, you will be exposed to the SDK and API, which will gradually make you aware of all of the capabilities and options you have to consider to build the high-quality apps.

Why the .NET version of ArcGIS Runtime?

You may be wondering why this book is focused on the .NET version of ArcGIS Runtime. Well, for starters, many Esri users like .NET, especially C#. Another reason is that if you know Java and Android, you'll find it easy enough to read the C# code because Java and C# are similar. Another important reason is that Esri has released 3D first in the .NET SDK, so it was hard to pass up this opportunity to talk about one of my favorite subjects (3D). ArcGIS Pro will also be based on .NET, so hopefully some of the skills you learn about here, especially **Model-View-ViewModel (MVVM)**, will also apply to Pro. Furthermore, there seems to be a general sense that Visual Studio is a great **Integrated Development Environment (IDE)**, as opposed to Eclipse, NetBeans, Google's Android Studio, and the other IDEs, but that's just my opinion. This is not anything against those IDEs, I just prefer Visual Studio. Very soon Esri is going to release the Xamarin® support, which means that the developers will be able to write an app in Visual Studio, compile it natively, and run it on Windows, Windows Store, iOS, and Android. I suspect you'll agree with some of these and disagree with others. Regardless, I hope that by the end of this book, you will have a solid understanding of the API so that you can move to any other SDKs, and it will be just a matter of switching your brain from C# to Android or to C++ or to Java.

System requirements for the ArcGIS Runtime SDK for .NET

In order to successfully install and use ArcGIS Runtime, you will need to have a PC that meets the minimum system requirements. You can find the system requirements on the ArcGIS Runtime site, `https://developers.arcgis.com/net/desktop/guide/system-requirements.htm`.

Setting up your environment

In order to start building an ArcGIS Runtime app, you will need the following IDEs and toolkits:

* Visual Studio 2015

* ArcGIS Runtime SDK for .NET 10.2.7

* MVVM Light

Visual Studio

First things first, we will be using Visual Studio 2015. You can download an evaluation version or the community version of Visual Studio 2015 for free from Microsoft. Here are the instructions for the community version:

1. Navigate to `https://www.visualstudio.com/products/free-developer-offers-vs.aspx`.

2. Click on **Download** under **Visual Studio Community**. This step will download a small installation program called `vs_community_ENU.exe`.

3. Run `vs_community_ENU.exe`.

4. Click on **I agree to the License Terms and Privacy Policy**.

5. Change the installation path if you want to, and decide whether you want to join the VS Experience Improvement Program.

6. Click on **Next**.

7. Click on the optional features. You can uncheck Microsoft Server data tools, Microsoft Web Developer Tools, and Silverlight Development Kit.

8. Click on **Install**.

The installation will take several minutes to complete. With the community version, you will be prompted to create an account. Follow the instructions. You will also be asked to create a Visual Studio Online site, and optionally, whether you want to create a Team Foundation Server or Git site.

ArcGIS Runtime for .NET SDK

In this book, we will be using version 10.2.7. To create an ArcGIS Runtime app, you will need to perform the following steps:

1. Navigate to the site `https://developers.arcgis.com/net/`.

2. Click on **Sign In**.

3. Below the **Sign In** textbox, you will see a link named **Sign up for free.**, as shown in the following screenshot.

4. Fill in your name and e-mail address. An e-mail will be sent to your account. Follow the instructions in the e-mail.

5. Once you've followed all of the instructions, you will now be able to download ArcGIS Runtime.

6. Navigate back to `https://developers.arcgis.com/net/`, and click on the large **Download SDK** button:

7. Once the download is complete, you can run the installation program. First, you need to tell it where to unzip to. In my case, C:\DevTools\ArcGIS 10.2.7 was chosen, as shown in this screenshot:

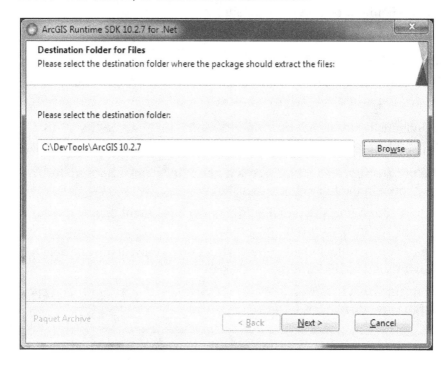

8. Next, you will be taken to another screen once the extraction has finished. Click on **Close**. Make sure **Launch the setup program** is checked.

9. On the **Welcome** screen, click on **Next >**. You will be taken to the **License Agreement** screen.

10. Click on **I accept the license agreement**, and then click on **Next >**.

11. You will then be taken to the **Select Features** screen, as shown in the following screenshot. Click on **Change...** to change the path if you wish. I've chosen C:\DevTools. Click on **Next >**:

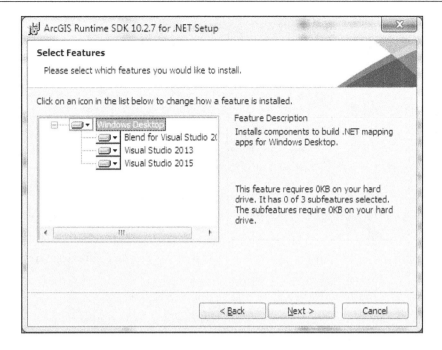

12. Now you will be taken to the **Ready to Install Program** page. Click on **Install**. This step will take several minutes.

If version 10.2.7 is no longer available on Esri's site, you can use NuGet to install it from Visual Studio with the NuGet Package Manager Console, as shown here:

```
Install-Package Esri.ArcGISRuntime –Version 10.2.7.1234
```

For more details, refer to `https://www.nuget.org/packages/Esri.ArcGISRuntime/`.

MVVM Light

Now that you have Visual Studio installed, you need to install MVVM Light. MVVM Light will be explained in the next chapter. Follow these steps:

1. Close Visual Studio if you have it open.

2. Navigate to `https://mvvmlight.codeplex.com/` and click on the **download** tab.

3. Download MVVM Light for VS 2015. This toolkit is a Visual Studio extension that will allow you to use MVVM light in Visual Studio. The extension of the file is `vsix`.

4. Run it by double-clicking on it.

5. Once you run the installer you will be prompted to install it. Click on **Install**.

6. Click on **Close** when completed.

Sample code

Throughout this book, we're going to build many apps to illustrate the concepts using step-by-step instructions. To aid you in performing these steps, this book comes with sample apps, which you can download from GitHub by following these steps:

1. Navigate to `https://github.com/rvinc66/ArcGISRuntimBook`.

2. Click on **Download ZIP** on the right-hand side of the page to download the source code.

3. Unzip the file to your `C:\` drive. This step will create a directory named `ArcGISRuntime-master`.

4. Simply rename this directory `ArcGISRuntimeBook`.

5. Inside the `ArcGISRuntimeBook` folder, you'll find a solution file (`ArcGISRuntimeBook.sln`).

This solution contains all the projects used in this book. Throughout the rest of this book, you'll find projects with chapter names, such as `Chapter2`, and `Chapter2A`. You can either follow the steps yourself to build your own projects or open a project that comes with these samples in the code you just downloaded. To run a particular sample, right-click on it and click on **Set As Startup Project**, and then run it.

Let's build an app!

Now that your development environment is established, your next task is to create an initial app.

The old way of building an app

In the following steps, we're going to create an app that operates like a traditional Windows Forms app. It will contain a UI and code-behind files just like those apps used to. Follow these steps:

1. Open Visual Studio 2015.

2. On the **Start Page**, click on **New Project...**. Refer to the following screenshot.

3. Under **Templates** on the left-hand side of the dialog, navigate to **Templates | Visual C# | Windows | Classic Desktop**.

4. Scroll to **ArcGIS Runtime 10.2.7 for .NET Application**.

5. In the **Name:** field, change the name to anything you want.

6. Click on **Browse…** beside the **Location:** field and specify the place where you want to place the project. You'll note that I've specified C:\AppDev\VS2015.

7. Click on **OK**:

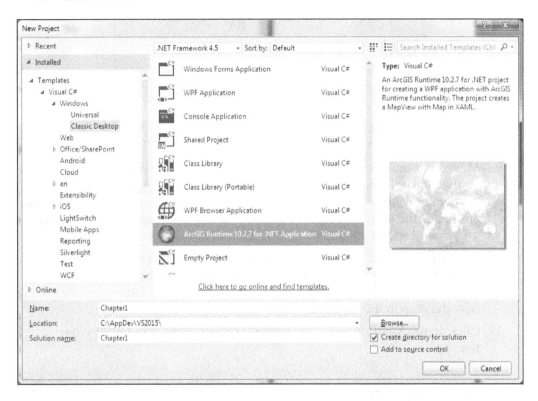

8. You now have an app. Click on **Start** on the main toolbar. Depending on your Internet connection speed, you should see a map as shown here:

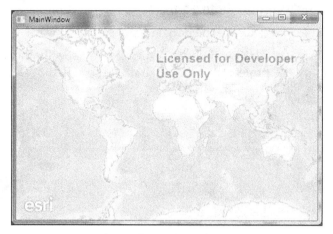

View of map after the functioning of Start

9. Move your mouse over the map and scroll forward to zoom in. As you zoom in, the map will become more detailed. Scroll backwards with your mouse to zoom out. Also, left-click on your mouse to pan around the map. You can also make the window larger or click on **Maximize** to make it fill the screen. You should spend some time getting used to how to navigate this map.

10. Another thing that you can do if you want to zoom to a specific area is hold down the *Shift* key and with your left mouse button drag a box around an area on the world map. We'll cover more of this subject later when we go over the map in more detail.

11. Click on the **X** button in the upper-right corner of the window to close the app.

Under the hood

Now that you've created a map and interacted with it, let's look under the hood at the code so that we can see how exactly this works. Follow these steps:

1. In the **Solution Explorer** pane, you will see that your current code looks like this screenshot:

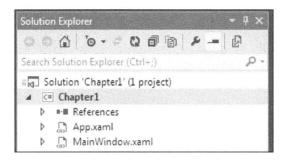

2. Double-click on `MainWindow.xaml`. From there, you will be shown the XAML and the design canvas for the XAML.

See the following code sample. In this XAML code, you will note that there is a Window element, typical XAML namespaces, title, height, and width. You will also note that there is an extra namespace called `xmlns:esri`, which is the default namespace for ArcGIS Runtime. This namespace is required to create the map. Every app you make throughout the rest of the book will contain this namespace:

```xml
<Window x:Class="Chapter1.MainWindow"
  xmlns="http://schemas.microsoft.com/winfx/2006/xaml/
    presentation"
  xmlns:x="http://schemas.microsoft.com/winfx/2006/xaml"
  xmlns:esri="http://schemas.esri.com/arcgis/runtime/2013"
    Title="MainWindow"
    Height="350"
    Width="525">
  <Grid>
    <esri:MapView x:Name="MyMapView"
      LayerLoaded="MyMapView_LayerLoaded">
    <esri:Map>
      <esri:ArcGISTiledMapServiceLayer ID="Basemap"

      ServiceUri="http://services.arcgisonline.com/ArcGIS/rest/
        services/World_Topo_Map/MapServer"/>
      </esri:Map>
    </esri:MapView>
  </Grid>
</Window>
```

The next thing you will note is that there is a Grid tag that contains a MapView class. This MapView class is a container for a map and it is a control. Note that the MapView class is prefixed with the esri prefix. This prefix tells Visual Studio where to find the object in the API. Next, note that in the MapView class, there is a map that is also prefixed with the esri namespace. Like any other XAML element, you will note that each element has the beginning <esri:Map> and the ending </esri:Map>. Between these tags, you will note that there is a layer that has been defined. It's a tiled Basemap layer, which has an ID, and there is a service **Uniform Resource Identifier (URI)** that defines where the map is coming from. As you can see from the URI, this layer is coming from AGOL, which is in the cloud. We'll discuss layers in more detail in the later chapters.

Now that you've seen the XAML, let's look at the code-behind file. Go back to the **Solution Explorer** pane, right-click on MainWindow.xaml, and click on **View Code**:

```
using Esri.ArcGISRuntime.Controls;
using System;
using System.Diagnostics;
using System.Linq;
using System.Windows;

namespace ArcGISApp1
{
    public partial class MainWindow : Window
    {
        public MainWindow()
        {
            InitializeComponent();
        }

        private void MyMapView_LayerLoaded(object sender,
            LayerLoadedEventArgs e)
        {
            if (e.LoadError == null)
                return;

            Debug.WriteLine(string.Format("Error while loading
                layer
                : {0} - {1}",
                e.Layer.ID, e.LoadError.Message));
        }
    }
}
```

Just like in a typical C# code, there are the using statements, a namespace declaration, a class for the Window element, a constructor, and an event handler. Most of this is pretty standard C# code, but the one thing worth mentioning is that in the XAML, note that there was an event named LayerLoaded, which is called when the basemap finished loading. All this code does is check whether e.LoadError equals null. If it does equal null, the event is returned. If it is not equal to null, an error message will be written in the **Output** window in Visual Studio.

The interesting thing about this code is that it is quite similar to the days of Windows Forms apps, which has a UI and code-behind file.

Enhancing the app

Let's add a couple of features to this app. First of all, let's add the scale to the map. After all, this is a mapping app and scale is important in any map. Perform the following steps:

1. Add the following lines shown in the code sample after the closing </esri:MapView> tag:

```
<TextBlock Background="#77000000"
   HorizontalAlignment="Center"
   VerticalAlignment="Bottom" Padding="5"
   Foreground="White">
     <Run>Map Scale 1:</Run>
     <TextBlock Text="{Binding ElementName=MyMapView,
        StringFormat={}{0:0}, Path=Scale}"
        FontSize="12" FontWeight="Bold"
   Foreground="White"/>
</TextBlock>
```

2. Run the app. You will see the scale shown at the bottom of the page, as shown in the following screenshot:

The way these steps work is actually just like any other **Windows Presentation Framework (WPF)** app. In this case, we're making a TextBlock tag bind to the MapView class. Check out the Text property of the inner TextBlock tag. It is being bound to the element called MyMapView. The MyMapView element is the identifier of the MapView class. It is also using the Scale property with Path from the MyMapView element.

In other words, the `MapView` class has a property called `Scale`, and the `TextBlock` tag is simply binding to it just like it's possible to bind to any other WPF `FrameworkElement` type. All other properties of the `TextBlock` tags are standard properties. There is also a `Run` element in the code just to hold the text, `Map Scale 1`. Another interesting thing you can glean from this is that if you want to place `FrameworkElement` types on top of the map, all you have to do is to place your XAML after the closing `MapView` element.

Of course, just showing a basemap with a scale isn't very interesting, so let's add some more layers that will allow us to actually do something, such as search for a state or a city. In order to do that, you will need to take a look at some of the publicly available map services on AGOL. Perform the following steps:

1. Navigate to `http://sampleserver6.arcgisonline.com/arcgis/rest/ services/USA/MapServer`.

2. The name of the service is USA. Click on **ArcGIS JavaScript** beside **View In**. You should be able to see a map of the United States.

3. Read the **Service Description** option. It's short in this case but tells you exactly what the service is all about.

4. Under **Layers**, you will note that there are four layers listed — **Cities**, **Highways**, **States**, and **Counties**.

5. Go back to Visual Studio and add the following lines of code to your XAML after the first layer. Make sure it is after the first layer and before the closing `</esri:Map>` tag:

```
<esri:ArcGISDynamicMapServiceLayer ID="USA"
   ServiceUri="http://sampleserver6.arcgisonline.com/arcgis/
      rest/services/USA/MapServer"/>
```

6. As you can see, a dynamic map service has been added using `ArcGISDynamicMapServiceLayer`. It was given a name, USA. A dynamic map service contains any number of layers. They return the geometry and attributes in the layer across the Internet in the form of an image.

7. Run the app and you will see a map with all the layers over USA. Feel free to zoom in so that you can see more details.

8. Close the app. The next thing that would make the app more interesting is to add the ability to search for a city, county, or state name; so let's do that.

9. Enter the XAML code, as shown here, after the closing `MapView` element:

```
<TextBlock Name="Search" Background="#77000000"
   HorizontalAlignment="Center"
   VerticalAlignment="Top" Padding="5"
   Foreground="White" >
```

```
<Run>Search for  </Run>
<TextBox Name="SearchTextBox"
  Text="Lancaster">
</TextBox>
<Run>  in the Cities, Counties or States layer. </Run>
<Button Content="Find" Width="30"></Button>
</TextBlock>
```

10. As you can see, a new `TextBlock` tag has been added after the closing `MapView` element with some text using `<Run>`, and a `Button` tag has been added.

11. Run your app. You will see the new XAML at the top of your map, as shown in the following screenshot. You can also see what it looks like in the XAML Designer workspace:

12. If you run this app and click on **Find**, nothing will happen because we haven't wired up the button to an event handler.

13. If you haven't done this before, read carefully. In the button's XAML after `Width="30"`, start typing the word `Click`, and then press the *Tab* key twice. You will note that as with other WPF apps, this action prompts the creation of an event handler, as shown here:

14. After you've done this step, an event handler will be created in `MainWindow.xaml.cs`. We will return to this event handler after the next step.

15. Let's make a few more changes to this code. First, in the `Window` element's properties, change the `Height` attribute to `600` and the `Width` attribute to `800`. Then, add some `Grid` row definitions, as shown in following code sample. Lastly, below the `TextBlock` tag, add a `DataGrid` tag and name it `MyDataGrid`. Make sure you've entered exactly what is shown in the code sample. Also, note that the scale related to XAML has been removed:

```
<Window x:Class="ArcGISApp1.MainWindow"
  xmlns="http://schemas.microsoft.com/winfx/2006/xaml/
    presentation"
```

```
    xmlns:x="http://schemas.microsoft.com/winfx/2006/xaml"
    xmlns:esri="http://schemas.esri.com/arcgis/runtime/2013"
    Title="MainWindow"
    Height="600"
    Width="800"
>
    <Grid>
      <Grid.RowDefinitions>
        <RowDefinition Height="400" />
        <RowDefinition Height="200" />
      </Grid.RowDefinitions>

      <esri:MapView x:Name="MyMapView" Grid.Row="0"
        LayerLoaded="MyMapView_LayerLoaded" >
        <esri:Map>
          <esri:ArcGISTiledMapServiceLayer ID="Basemap"
            ServiceUri="http://services.arcgisonline.com/
              ArcGIS/rest/services/World_Topo_Map/
                MapServer"/>
          <esri:ArcGISDynamicMapServiceLayer ID="USA"
            ServiceUri="http://
              sampleserver6.arcgisonline.com/arcgis/rest/
                services/USA/MapServer"/>
        </esri:Map>
      </esri:MapView>

      <TextBlock Grid.Row="0" Name="Search"
        Background="#77000000"
        HorizontalAlignment="Center"
        VerticalAlignment="Top" Padding="5"
          Foreground="White" >
        <Run>Search for  </Run>
        <TextBox Name="SearchTextBox"
          Text="Lancaster"></TextBox>
        <Run>  in the Cities, Counties or States layer:
          </Run>
        <Button Content="Find" Width="30"
          Click="Button_Click"></Button>
      </TextBlock>

      <DataGrid Name="MyDataGrid" Grid.Row="2"
        Height="200"></DataGrid>

    </Grid>
</Window>
```

16. Right-click on `MainWindow.xaml` and click on **View Code**, or just double-click on it.

17. Add the following `using` statements at the top of the source file:

```
using Esri.ArcGISRuntime.Controls;
using Esri.ArcGISRuntime.Data;
using Esri.ArcGISRuntime.Geometry;
using Esri.ArcGISRuntime.Layers;
using Esri.ArcGISRuntime.Tasks;
using Esri.ArcGISRuntime.Tasks.Query;
```

18. Also, note that in the source code file, there is a new `private` method called `Button_Click`. It currently has no code in it. We will now populate it with some code so that we can search for a city, county, or state. Enter the following code in the event handler (`Button_Click`). Also note that the `async` keyword has been added to the event's signature:

```
private async void Button_Click(object sender,
    RoutedEventArgs e)
{

    var url = "http://sampleserver6.arcgisonline.com/
        arcgis/rest/services/USA/MapServer";
    var findTask = new FindTask(new Uri(url));

    var findParameters = new FindParameters();
    findParameters.LayerIDs.Add(0); // Cities
    findParameters.LayerIDs.Add(3); // Counties
    findParameters.LayerIDs.Add(2); // States

    findParameters.SearchFields.Add("name");
    findParameters.SearchFields.Add("areaname");
    findParameters.SearchFields.Add("state_name");

    findParameters.ReturnGeometry = true;
    findParameters.SpatialReference =
        MyMapView.SpatialReference;

    findParameters.SearchText = SearchTextBox.Text;
    findParameters.Contains = true;

    FindResult findResult = await
        findTask.ExecuteAsync(findParameters);

    var foundCities = 0;
    var foundCounties = 0;
```

```
      var foundStates = 0;

      // Loop through results; count the matches found in
          each layer
      foreach (FindItem findItem in findResult.Results)
      {
          switch (findItem.LayerID)
          {
              case 0: // Cities
                  foundCities++;
                  break;
              case 3: // Counties
                  foundCounties++;
                  break;
              case 2: // States
                  foundStates++;
                  break;
          }
      }

      // Report the number of matches for each layer
      var msg = string.Format("Found {0} cities, {1}
          counties, and
          {2} states containing '" + SearchTextBox.Text +
          "' in a
          Name attribute", foundCities, foundCounties,
      foundStates);
      MessageBox.Show(msg);



      // Bind the results to a DataGrid control on the
          page
      MyDataGrid.ItemsSource = findResult.Results;
  }
```

19. Once you've completed entering the code, run the app. You should note that the window is larger and that there is a data grid at the bottom of the window. You will also see all the layers in the dynamic map service. You can now click on the **Find** button. A message box will appear, indicating the number of cities, counties, and states that have the name Lancaster in them. Refer to the included sample project code with this book, named Chapter1, if you need any help:

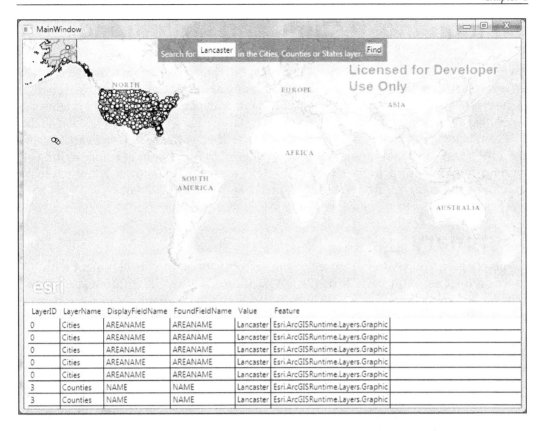

LayerID	LayerName	DisplayFieldName	FoundFieldName	Value	Feature
0	Cities	AREANAME	AREANAME	Lancaster	Esri.ArcGISRuntime.Layers.Graphic
0	Cities	AREANAME	AREANAME	Lancaster	Esri.ArcGISRuntime.Layers.Graphic
0	Cities	AREANAME	AREANAME	Lancaster	Esri.ArcGISRuntime.Layers.Graphic
0	Cities	AREANAME	AREANAME	Lancaster	Esri.ArcGISRuntime.Layers.Graphic
0	Cities	AREANAME	AREANAME	Lancaster	Esri.ArcGISRuntime.Layers.Graphic
3	Counties	NAME	NAME	Lancaster	Esri.ArcGISRuntime.Layers.Graphic
3	Counties	NAME	NAME	Lancaster	Esri.ArcGISRuntime.Layers.Graphic

That was the name of my home town. Try yours. Note that not all cities in the USA are in the `Cities` layers.

There were several things in the code-behind file. First, the URL of the map service was set, a `FindTask` constructor was created, a `FindParameters` argument was added, search fields were set, a parameter was set to return the geometry, the spatial reference was set, the search text was set, a parameter was set as to whether to use the exact text entered, `FindResult` was executed asynchronously, the results were looped over to find the counts in each layer, a message was created with the counts, and finally, and most importantly, the results were passed into the `DataGrid` tag's `ItemsSource` so that it can be viewed. In the next chapters, we will go over these objects in more detail.

Configuring, testing, and structuring

Now that you've created a simple app, similar to a traditional Windows Forms app, except with WPF, we're going to go into the next chapter and explain why this is not an ideal approach. We're going to redesign this app so that it's more reusable. What is the best way to write code that is reusable, maintainable, and configurable? How would we automate the testing of this app as it stands? How would you make it so that you could change the base layer without having to change the XAML and recompile the app? These are all important questions that need to be thought of before writing a professional level app. In the chapters to come, we will address these issues while explaining the main features of ArcGIS Runtime in much more detail.

Summary

In this chapter, you've been exposed to the full breadth of app development with Esri's premier native apps SDK. You were introduced to ArcGIS Runtime, provided with high-level examples of its use and application, the language choices you have, a little history of Esri's development tools, the use of native apps across the broader app market, the types of mapping apps, the architecture of ArcGIS Runtime, areas where ArcGIS Runtime fits within the Esri platform, installation steps, building your first app with ArcGIS Runtime, and finally, some things to think about when it comes to configuration, testing, and structuring your app.

In the next chapter, we're going to take the app we've built here and refactor it so that it fits the MVVM pattern. This will allow us to make our apps maintainable and provide us with a structure for the remaining chapters.

2
The MVVM Pattern

In this chapter, we're going to address the topic of structuring code for a WPF app. We are going to discuss a pattern called MVVM, which is a standard pattern Microsoft has developed for WPF apps. We will also learn to configure the app so that it's easy to change important information such as search parameters and the layers that it uses.

The following topics will be discussed:

- MVVM
- Configuration

Model-View-ViewModel

MVVM is a pattern developed by Microsoft to primarily support WPF-based apps. It applies to Windows, Windows Store, and Windows Phone. The primary goals of the pattern are as follows:

- This pattern supports **Separation of Concerns (SoC)**. One of the biggest problems of writing code for the old Windows Forms app was that the entire code existed directly behind the UI. As a result, one change could result in problems across the rest of the app. MVVM changes this pattern by separating sections of the code into distinct parts. In particular, the application logic, data, and UI are separated. This allows for new specialties, such as that UI developers can create new and interesting UIs while someone else can work on the application logic.

- As XAML supports binding, commands, and dependency properties, the MVVM pattern is ideal because it allows the developer to connect everything together so that the app can be reskinned (by changing out the UI) without breaking the application logic.

- MVVM makes your apps more easily maintained. If there's an error in the logic at one place, fixing the error fixes the problem everywhere else.

- Another key advantage of MVVM is that it makes apps more testable because the application is separated from the UI, which means that the application logic can be tested in some cases without a UI.

The pattern

As we can see in this diagram, there are three distinct levels in MVVM:

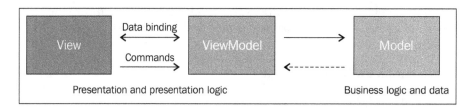

The **Model** block is at the lowest level of the pattern. It represents your data in the form of a .NET class. However, it is not the persistence layer of the app; it's just a data representation in the form of properties, for the most part. Populating the Model block is typically taken care of by some kind of service.

The **View** block is the XAML or UI of your app. In the previous chapter, it was the `MainWindow.xaml` file. As shown in the *Let's build an app* example in the section in *Chapter 1, Introduction to ArcGIS Runtime*, it shows the data and allows for user interaction. As you saw in that example, there was a lot of code in the `MainWindow.xaml.cs` file and this is not what you want because the code is directly behind the View. With MVVM, the ideal goal is for no code-behind files. Note that it is ideal, but sometimes not practical or warranted, depending on your app's requirements and the time frame in which you have to develop the app.

The **ViewModel** block is the guts of your app. It contains all application business logic. It is really an abstraction layer between the **Model** and **View** block. Firstly, it pulls in data from the **Model** block. Secondly, it handles interaction from the user by exposing methods and commands that the **View** block binds to. In summary, the **ViewModel** block is an intermediary between the **View** block and the **Model** block. It should be lightweight and rely on other classes to do the real work.

If you look back at the diagram, you'll note that there is a two-way interaction between the **View** and **ViewModel** blocks and a two-way interaction between the **ViewModel** and **Model** blocks. The key enabler of MVVM is data binding. It really is the glue for MVVM and WFP in general. Binding has become popular, even in web development. For example, AngularJS relies on binding, although it uses a different pattern.

When it comes to WPF's features, they can be divided up across the MVVM pattern, as shown in this diagram:

This diagram makes sense because bindings, behaviors, animations, themes, windows, pages, and user controls are all UI elements. For the **ViewModel** level, properties and commands are just the typical .NET types. Lastly, the data, domain logic, and class properties lie in the **Model** tier.

MVVM in action (without a toolkit)

Setting up an app to use the MVVM pattern is relatively straightforward, but can involve a lot of boiler plate code. Let's illustrate this so that you can see all of the steps and the way things work exactly:

1. Create a new ArcGIS Runtime app as you did in *Chapter 1, Introduction to ArcGIS Runtime*. Name your project Chapter2.

2. Go ahead and copy the XAML and code-behind file from that project to this new project.

3. Create a new, standard C# class. Name the file `Model.cs`, as shown here:

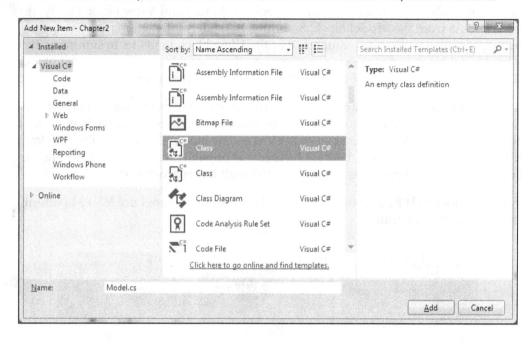

4. Add the following code to the `Model` class:

```csharp
using System;
using System.Collections.Generic;
using System.Linq;
using System.Text;
using System.Threading.Tasks;

namespace Chapter2
{
    public class Model
    {
        private string searchText = "Lancaster";
        private string basemapLayerUri =
            "http://services.arcgisonline.com/ArcGIS/rest/
                services/World_Topo_Map/MapServer";
        private string usaLayerUri = "http://
            sampleserver6.arcgisonline.com/arcgis/rest/
                services/USA/MapServer";

        public Model()  { }

        public string SearchText
        {
            get { return this.searchText; }
```

```
        set
        {
            if (value != this.searchText)
            {
                this.searchText = value;
            }
        }
    }
    public string BasemapLayerUri
    {
        get {return this.basemapLayerUri; }
        set
        {
            if (value != this.basemapLayerUri)
            {
                this.basemapLayerUri = value;
            }
        }
    }
    public string USALayerUri
    {
        get  { return this.usaLayerUri; }
        set
        {
            if (value != this.usaLayerUri)
            {
                this.usaLayerUri = value;
            }
        }
    }
  }
}
```

Most of this is a pretty basic .NET code. It simply stores our layer's URIs and a search string. Now that the Model class has been created, it's time to create the ViewModel class:

1. Add references to the `System.ComponentModel` and `System.Runtime.CompilerServices` namespaces.

2. Add another class (name it `ViewModel.cs`) to the project and enter this code:

```
using System;
using System.Collections.Generic;
using System.Linq;
using System.Text;
using System.Threading.Tasks;
using System.Windows.Input;
using System.ComponentModel;
using System.Runtime.CompilerServices;
```

```csharp
namespace Chapter2
{
    public class ViewModel : INotifyPropertyChanged
    {
        public Model myModel { get; set; }

        public event PropertyChangedEventHandler
            PropertyChanged;

        public ViewModel()
        {
            myModel = new Model();
        }

        public string BasemapUri
        {
            get { return myModel.BasemapLayerUri;  }
            set
            {
                this.myModel.BasemapLayerUri = value;
                OnPropertyChanged("BasemapUri");
            }
        }
        public string USAUri
        {
            get { return myModel.USALayerUri; }
            set
            {
                this.myModel.USALayerUri = value;
                OnPropertyChanged("USAUri");
            }
        }
        public string SearchText
        {
            get { return myModel.SearchText; }
            set
            {
                this.myModel.SearchText = value;
                OnPropertyChanged("SearchText");
            }
        }
```

```
protected void OnPropertyChanged([CallerMemberName]
    string
    member = "")
{
    var eventHandler = PropertyChanged;
    if (eventHandler != null)
    {
        PropertyChanged(this, new
                PropertyChangedEventArgs(member));
    }
}
}
}
```

Our `ViewModel` class looks a little more involved. It contains an event (`PropertyChangedEventHandler`) and method (`OnPropertyChanged`). The `OnPropertyChanged` method is really the method that makes binding work in WPF apps. When this method is called, a WFP element updates itself based on firing an event, such as a text change in a textbox. The `CallerMemberName` attribute will pass the property name that executed `OnPropertyChanged`. Whenever a property is changed, `OnPropertyChanged` is called.

3. Open `MainWindow.xaml` and add the following line along with the other namespaces:

```
xmlns:local="clr-namespace:Chapter2"
```

Note that your namespace will be different if you used a different project name.

4. Add the following resource for `Window`. This resource informs `Window` to set its `DataContext` class in the `ViewModel` class and use it throughout the entire `Window` resource:

```
<Window.Resources>
    <local:ViewModel x:Key="VM"/>
</Window.Resources>
```

These lines of code instruct the View class to use the `ViewModel` class and name it `VM`.

5. Next, add the `DataContext` class to the `Grid` tag, which is the root element:

```
<Grid DataContext="{StaticResource VM}">
```

This line of code is pretty powerful. It's basically telling the View (XAML) class to use this `ViewModel` class as the data source. In reality, the `ViewModel` class is actually relying on the `Model` class to handle the data, but the `DataContext` class of View has no idea that this is occurring (remember SoC?).

6. The next task is to update the layer binding to use the properties on the `ViewModel` class. Make the changes as in the following code:

```
<esri:ArcGISTiledMapServiceLayer ID="Basemap"
    ServiceUri="{Binding Source={StaticResource VM},
    Path=BasemapUri}"/>
<esri:ArcGISDynamicMapServiceLayer ID="USA"
    ServiceUri="{Binding Source={StaticResource VM},
    Path=USAUri}"/>
```

7. Also, add this line:

```
<TextBox Name="SearchTextBox" Text="{Binding
    SearchText}"></TextBox>
```

These changes bind the elements to the properties of `ViewModel`. Also, two ways of binding to the `ViewModel` class have been shown. The `ServiceUri` properties are referring to the `ViewModel` class, while `SearchTextBox` is just referring to the property without specifying the `ViewModel` class. This is possible because the property is looking up to the root element, which is our `ViewModel` class.

The XAML code is shown here in its entirety:

```
<Window x:Class="Chapter2.MainWindow"
    xmlns="http://schemas.microsoft.com/winfx/2006/xaml/
        presentation"
    xmlns:x="http://schemas.microsoft.com/winfx/2006/xaml"
    xmlns:esri="http://schemas.esri.com/arcgis/runtime/
        2013"
    xmlns:local="clr-namespace:Chapter2"
        Title="MainWindow"
        Height="600"
        Width="800">

    <Window.Resources>
        <local:ViewModel x:Key="VM"/>
```

```xml
        </Window.Resources>

        <Grid DataContext="{StaticResource VM}">
            <Grid.RowDefinitions>
                <RowDefinition Height="*" />
                <RowDefinition Height="Auto" />
            </Grid.RowDefinitions>

            <esri:MapView x:Name="MyMapView" Grid.Row="0"
                LayerLoaded="MyMapView_LayerLoaded" >
                <esri:Map>
                    <esri:ArcGISTiledMapServiceLayer
                        ID="Basemap"
                        ServiceUri="{Binding
                        Source={StaticResource VM},
                        Path=BasemapUri}"/>
                    <esri:ArcGISDynamicMapServiceLayer ID="USA"
                        ServiceUri="{Binding
                            Source={StaticResource VM},
                            Path=USAUri}"/>
                </esri:Map>
            </esri:MapView>

            <TextBlock Grid.Row="0" Name="Search"
                Background="#77000000"
                HorizontalAlignment="Center"
                VerticalAlignment="Top" Padding="5"
                Foreground="White" >
                <Run>Search for  </Run>
                <TextBox Name="SearchTextBox"
                    Text="{Binding SearchText}">
                </TextBox>
                <Run>  in the Cities, Counties or States layer.
                    </Run>
                <Button Content="Find" Width="30"
                    Click="Button_Click">
                </Button>
            </TextBlock>

            <DataGrid Name="MyDataGrid" Grid.Row="2"
                Height="200" ></DataGrid>

        </Grid>
    </Window>
```

8. Run the app and you will see the same layers as you saw in *Chapter 1, Introduction to ArcGIS Runtime*. If you don't, find the code to be as shown here and resolve any errors, and if you need any help following along, see the sample project included with this book; it's named `Chapter2`.

So, with a few changes, we satisfied SoC to a certain degree. We now have `Model`, `View`, and `ViewModel` classes. While this may seem like a lot of changes to achieve the same app which we built in the previous chapter, you will find that as your apps become more complex, this pattern will allow multiple people to work on the app. And, as we will see, this pattern allows us to test our app.

While this is a huge improvement, there is actually a lot of boilerplate code. Imagine if your app had 20 Models, 10 ViewModels, and several UIs. We didn't even deal with the button here because it would have resulted in even more boilerplate code. In effect, you'd end up repeating a lot of this code, such as `OnPropertyChanged` and `PropertyChangedEventHandler`. You could roll a base class to handle this of course, and that would help, but the good news is that this has been dealt with by others. Enter MVVM Light.

MVVM Light to the rescue

If you search for MVVM in your favorite search engine, you will find that several MVVM libraries are available. Microsoft originally developed one library called **Prism**. Since then, several other libraries have come along, such as Unity, Caliburn, Micro, and MVVM Light. It is highly recommended that you evaluate all of them to find one that suits your team and project. In this book, we're going to use MVVM Light because, as the name implies, it's a lightweight version of the full implementation of the MVVM pattern. It also can be applied to Windows Store and Windows Phone.

Let's rebuild the app again, but this time with MVVM Light. Create a new ArcGIS Runtime app as you've done before. In the following steps, the app is called `Chapter2a`:

1. Using NuGet, install MVVM Light in your new project. Right-click on the project and click on **Manage NuGet Packages**. In the search box in the upper-right corner of the window, type mvvmlight, as shown here:

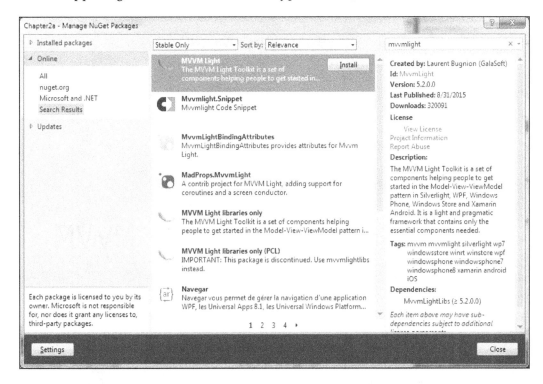

2. Click on **MVVM Light**, and then click on **Install**. All of the necessary references will be added. Note that we are using Version 5.2. Click on **Close**.

3. Create a new directory in your project and call it ViewModels.

4. Add an MVVM Light `ViewModel` class to your project. Right-click on the `ViewModels` directory and add a new item, and then find **MvvmViewModel (WPF)**. It will be named `MainViewModel.cs` in these examples, as shown here:

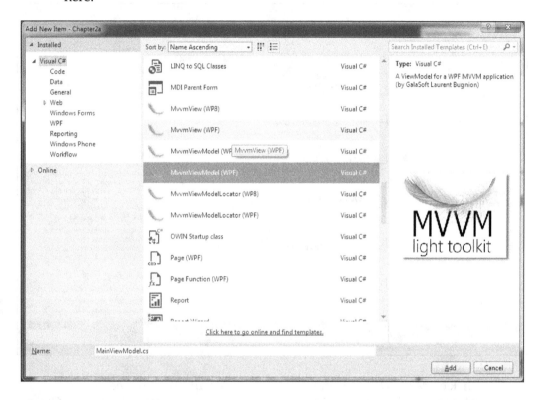

5. Right-click on the `ViewModels` directory and add a new item. Click on **MvvmViewModelLocator (WPF)**. It will be named `ViewModelLocator` here:

A `ViewModelLocator` class simply allows MVVM Light to find your `ViewModel` classes and execute them for you. This is called **Inversion of Control (IoC)**. IoC is a powerful technique that helps you manage and find `ViewModels`. Open `ViewModelLocator`. The first thing you will see is an example of how to use it as an application resource, and how to use it as a `DataContext` class in the `View` class. In the constructor, you'll note `ServiceLocator.SetLocatorProvider`.... This is a Microsoft class that lets your `ViewModel` classes be instantiated without any concern about the implementation of how this is done. After this, you will note that these lines are commented out:

```
////if (ViewModelBase.IsInDesignModeStatic)
////{
////    // Create design time view services and models
////    SimpleIoc.Default.Register<IDataService,
    DesignDataService>();
////}
////else
////{
////    // Create run time view services and models
////    SimpleIoc.Default.Register<IDataService, DataService>();
////}
```

This step is an important part of MVVM Light. It allows you set up your project with mock data so that you can view your XAML code in Microsoft Blend, and so you won't get errors in design mode. The other option is to load data from a data service. We're going to skip this step because we're building a custom mapping app.

The next code segment is a property that your `View` class will use to set the `DataContext` class in the `View` class:

```
public MainViewModel MainViewModel
{
    get
    {
        return ServiceLocator.Current.GetInstance
            <MainViewModel>();
    }
}
```

This `get` statement simply returns an instance of the `ViewModel` class called `MainViewModel` so that you don't have to instantiate it yourself. Lastly, there is a `Cleanup` method that allows you to clean up resources.

Now that you've been briefly introduced to IoC and `ServiceLocator`, let's put this into action and make the `View` class use this `ViewModel` class. Note that the `MainWindow` file is in the `View` class that will consume the `MainViewModel` item:

1. Copy these lines to `App.xaml`:

   ```
   <Application.Resources>
       <vm:ViewModelLocator x:Key="Locator"
       d:IsDataSource="True" />
   </Application.Resources>
   ```

2. In `MainWindow.xaml`, set the `DataContext` class of the `Grid` tag as shown here:

   ```
   <Grid DataContext="{Binding  Source={StaticResource
     Locator}, Path=MainViewModel}">
   ```

3. Run the app. You will note that nothing has really changed, but everything is now set up to start coding our app.

Let's complete this new app by implementing the properties and commands needed to have a fully functioning app. This time, however, we're going to simplify the app, so we can focus on the most important aspects of the things happening here.

1. Create a new folder and name it `Models`. Copy the `Model.cs` file from the previous example to your new project and place it in the `Models` folder. Make sure to change the namespace to `Chapter2a.Models`.

2. In `MainViewModel`, make the following changes. Make sure you have the `using` statements shown here:

```
using System.Collections.Generic;

using GalaSoft.MvvmLight;
using GalaSoft.MvvmLight.CommandWpf;

using Chapter2a.Models;

using Esri.ArcGISRuntime.Tasks.Query;
```

3. Add the following `private` members:

```
private Model myModel = null;
private IReadOnlyList<FindItem> listResults;
public RelayCommand SearchRelayCommand { get;
    private set; }
```

The `SearchRelayCommand` class is an MVVM Light feature which allows you to invoke a method from the View class. All that's necessary is that you define the command as we just did.

4. Change the constructor to the following code:

```
public MainViewModel()
{
    if (IsInDesignMode)
    {
        // Code runs in Blend --> create design
        time data.
    }
    else
    {
        // Code runs "for real"
        this.myModel = new Model();
        this.SearchRelayCommand = new
        RelayCommand(Search);
    }
}
```

Note that we've instantiated `RelayCommand` and set it to the `public` property we created in the previous step. Also, note that we passed in `Search`, which is the method we're going to create next.

5. Add the method shown in the following code:

```
public async void Search()
{
    var url = "http://sampleserver6.arcgisonline.com/
        arcgis/rest/services/USA/MapServer";
    var findTask = new FindTask(new System.Uri(url));

    var findParameters = new FindParameters();
    findParameters.LayerIDs.Add(0); // Cities
    findParameters.LayerIDs.Add(3); // Counties
    findParameters.LayerIDs.Add(2); // States

    findParameters.SearchFields.Add("name");
    findParameters.SearchFields.Add("areaname");
    findParameters.SearchFields.Add("state_name");

    findParameters.ReturnGeometry = true;
    //findParameters.SpatialReference =
        MyMapView.SpatialReference;

    findParameters.SearchText = this.SearchText;
    findParameters.Contains = true;

    FindResult findResult = await
    findTask.ExecuteAsync(findParameters);

    var foundCities = 0;
    var foundCounties = 0;
    var foundStates = 0;

    // Loop thru results; count the matches found in each
        layer
    foreach (FindItem findItem in findResult.Results)
    {
        switch (findItem.LayerID)
        {
            case 0: // Cities
                foundCities++;
                break;
            case 3: // Counties
                foundCounties++;
                break;
            case 2: // States
                foundStates++;
```

```
                    break;
            }
    }

    // Report the number of matches for each layer
    var msg = string.Format("Found {0} cities, {1}
        counties, and {2} states containing '" +
            this.SearchText +
        "' in a Name attribute", foundCities,
            foundCounties,
        foundStates);

    // Bind the results to a DataGrid control on the page
    IReadOnlyList<FindItem> temp = findResult.Results;

    ObservableCollection<FindItem> obsCollection = new
        ObservableCollection<FindItem>();
    foreach (FindItem item in temp)
    {
        obsCollection.Add(item);
    }

    this.GridDataResults = obsCollection;

    System.Diagnostics.Debug.WriteLine(msg);
}
```

6. Note that the `SpatialReference` class has been commented out for now, and that the method is modified with `async`. Also, note that we've replaced `SearchTextBox.Text` with `this.SearchText`. More important than anything else, we've removed the `Button_Click` event from `Window` to the `ViewModel` class, which greatly reduced the code-behind file. Also, we've taken the results from `findResult` and placed them in `ObservableCollection` so that the `DataGrid` control can bind to it via a `ViewModel` property. Lastly, note that the `MessageBox` component has been removed. We'll return to `MessageBox` issues later. We'll run this app in a few steps, so make sure to look at the **Output** window to see the results.

7. Lastly, add the following four properties to this `ViewModel` class:

```
public ObservableCollection<FindItem> GridDataResults
{
    get { return this.listResults; }
    set
    {
        this.listResults = value;
```

```
                RaisePropertyChanged("GridDataResults");
            }
        }
        public string SearchText
        {
            get { return this.myModel.SearchText; }
            set
            {
                this.myModel.SearchText = value;
                RaisePropertyChanged("SearchText");
            }
        }
        public string BasemapUri
        {
            get { return myModel.BasemapLayerUri; }
            set
            {
                this.myModel.BasemapLayerUri = value;
                RaisePropertyChanged("BasemapUri");
            }
        }
        public string USAUri
        {
            get { return myModel.USALayerUri; }
            set
            {
                this.myModel.USALayerUri = value;
                RaisePropertyChanged("USAUri");
            }
        }
```

Note that we didn't have to write any boilerplate code as we did earlier. We didn't have to add any references, such as `System.ComponentModel` and `System.Runtime.CompilerServices`, and we didn't have to add `INotifyPropertyChanged` to our `ViewModel` class. This has been taken care of for us by MVVM Light because the `ViewModel` class inherits from `ViewModelBase`, which takes care of everything for us. Also, take a look at the constructor in `MainViewModel` and you'll also note that there is code for running your app in design mode or "for real". As you continue to use MVVM Light, even more benefits will become apparent.

Now, let's make the `View` class call a method on the `ViewModel` class via the `Find` button:

1. Delete the `Grid` control in the XAML code and replace it with all of this code:

```
<Window x:Class="Chapter2a.MainWindow"
    xmlns="http://schemas.microsoft.com/winfx/2006/xaml/
        presentation"
    xmlns:x="http://schemas.microsoft.com/winfx/2006/xaml"
    xmlns:esri="http://schemas.esri.com/arcgis/runtime/
        2013"
    xmlns:d="http://schemas.microsoft.com/expression/blend/
        2008"
    xmlns:locator="clr-namespace:Chapter2a.ViewModels"
    Height="600"
    Width="800">

    <Grid DataContext="{Binding  Source={StaticResource
        Locator},
        Path=MainViewModel}">
        <Grid.RowDefinitions>
            <RowDefinition Height="*" />
            <RowDefinition Height="Auto" />
        </Grid.RowDefinitions>

        <esri:MapView x:Name="MyMapView" Grid.Row="0"
            LayerLoaded="MyMapView_LayerLoaded">
            <esri:Map>
                <esri:ArcGISTiledMapServiceLayer
                    ID="Basemap"
                    ServiceUri="{Binding
                        MainViewModel.BasemapUri,
                    Source={StaticResource Locator}}"/>
                <esri:ArcGISDynamicMapServiceLayer ID="USA"
                    ServiceUri="{Binding MainViewModel.USAUri,
                Source={StaticResource Locator}}"/>
            </esri:Map>
        </esri:MapView>

        <TextBlock  Name="Search" Background="#77000000"
            HorizontalAlignment="Center"
            VerticalAlignment="Top" Padding="5"
            Foreground="White" >
            <Run>Search for  </Run>
            <TextBox Name="SearchTextBox" Text="{Binding
                SearchText}"></TextBox>
```

```
                    <Run>  in the Cities, Counties or States layer.
                       </Run>

                    <Button Content="Find" Width="30"
                       Command="{Binding
                       SearchRelayCommand}" >
                    </Button>
                </TextBlock>

                <DataGrid Name="MyDataGrid"  ItemsSource="{Binding
                    Path=GridDataResults, Mode=TwoWay,
                    UpdateSourceTrigger=PropertyChanged}"
                    AutoGenerateColumns="True"
                    Grid.Row="2" Height="200" ></DataGrid>
            </Grid>
        </Window>
```

First, note the syntax of the layers that are being bound to `Locator` via a static resource, and now `Binding` is set to `MainViewModel.BasemapUri`. This is necessary because these elements aren't in `Grid` hierarchy; they are in `Locator`. As a result, we had to refer to the property of `Locator` called `MainViewModel`, which has a property called `BasemapUri`. Also, note that the button now uses a command so that when you click on it, it no longer goes to the button's event handler in the code-behind file; it now calls a command called `SearchRelayCommand`, which is a way to call a method in the `ViewModel` class. Pretty nice, right?

2. Run the app. Most things are pretty much the same. Click on **Find** and observe that the `DataGrid` control is updated but the message goes to the Visual Studio Output window. You will see the same kind of results, however. If you need help, refer to the sample code provided in the project named `Chapter2a`.

If we had done this without MVVM Light, we would have had to add another reference to `System.Windows.Interactivity` and this code inside the button, to make the button work:

```
<Button Content="Find" Width="30" >
    <i:Interaction.Triggers>
        <i:EventTrigger EventName="Click">
            <i:InvokeCommandAction Command="{Binding
```

```
            SearchRelayCommand}"/>
        </i:EventTrigger>
    </i:Interaction.Triggers>
</Button>
```

Interaction is an advanced feature of WPF and MVVM, so using MVVM Light makes this feature easier to use and maintain.

With that said, the major difference now is that we've satisfied SoC by using binding and commands, which separates the View, ViewModel, and Model classes. You could have three people working on their own respective parts. The visual designer could work on the View concept, and other developers could work on the Model and ViewModel concepts.

MVVM Light command parameters

Now that we've enhanced the app with MVVM Light, there are still some things missing that will be required in a production-level application. If you noted earlier, the line in the Search method that sets the SpatialReference class was commented out because the ViewModel class doesn't have a reference to the map's spatial reference. The spatial reference tells the map about the projection or coordinate system that it should use. While this wasn't a problem in this particular example, it will be a problem in other situations. There are a few approaches to take to resolve this problem:

- You can pass the SpatialReference ID with each command. That would certainly work, but you'll end up having to pass that every time for every command.

- You can add the map as an application-level resource, and then refer to it in your ViewModel classes, as in this code:

```
this.map = App.Current.Resources["theMap"] as Map;
```

  This solution would work too and it fits well with the intent of MVVM, for the most part.

- Another option is to add a custom behavior to Window so that when the Window page opens, it passes MapView to your ViewModel class. This also satisfies the notion of SoC and it leaves the XAML code in Windows (or UserControl) so that it's easy to view it in design mode.

The big takeaway here is that there are many ways to deal with this problem. We're going to go with the first option for now because it is easy to illustrate by way of passing the `SpatialReference` ID as a `command` parameter. In the next chapter, we'll tackle behaviors. Let's make a few changes to the XAML code and the `ViewModel` class:

1. In `ViewModel.cs` in the `Chapter2a` project, add a `using` statement to `Esri.ArcGISRuntime.Geometry`, and then change `RelayCommand` to use an integer. The reason for this is that we simply want to pass the map's `SpatialReference` ID from the `View` class to the `ViewModel` class. Every spatial reference has a unique ID. Look at this line:

    ```
    public RelayCommand SearchRelayCommand { get; private set;
        }
    ```

2. Change the preceding line to this:

    ```
    public RelayCommand<int> SearchRelayCommand { get; private
        set; }
    ```

3. Now, have a look at this line:

    ```
    this.SearchRelayCommand = new RelayCommand(Search);
    ```

4. Again, change the preceding line to this:

    ```
    this.SearchRelayCommand = new RelayCommand<int>(Search);
    ```

5. Now, modify this line:

    ```
    public async void Search()
    ```

6. The preceding line is to be modified to this line:

    ```
    public async void Search(int wkid)
    ```

7. As you will recall in the last section, we commented out this line of code:

    ```
    //findParameters.SpatialReference =
        MyMapView.SpatialReference;
    ```

8. Go ahead and remove the line, and add these lines in its place:

    ```
    SpatialReference sr = new SpatialReference(wkid);
        findParameters.SpatialReference = sr;
    ```

9. Open up `MainWindow.xaml` so that we can pass the map's `SpatialReference` ID as a `command` parameter and the `Search` method can use it. Now, look at the following line:

    ```
    <Button Content="Find" Width="30" Command="{Binding
        SearchRelayCommand}" >
    </Button>
    ```

10. Change the preceding line to this line:

```
<Button Content="Find" Width="30" Command="{Binding
    SearchRelayCommand}" CommandParameter="{Binding
    Path=SpatialReference.Wkid,  ElementName=MyMapView}" >
</Button>
```

11. Run the app and click on **Find**. Nothing will be changed, but if you put a `break` point in the `Search` method and inspect the `wkid`, you'll see that it has an ID value of `102100`. This is the `SpatialReference` ID of a **Mercator** projection, which just happens to be the projection of the basemap. This works because we've told the XAML code to pass the `SpatialReference` ID from `MapView` (ID is `MyMapView`) using a `command` parameter.

You have now successfully passed in the information that the `Search` method really needs to operate correctly by passing information from the `View` class to the `ViewModel` class. More importantly, this doesn't break SoC. We could have passed `MapView` to the `Search` method but that would have put a UI control, in this case ArcGIS Runtime, in our `ViewModel` class and we don't want that.

Services and configuration

As you saw in the previous section, we instantiated the `Model` class directly in the `ViewModel` class. This was fine for a simple app, but for a production-level app, you are going to want to create services that take care of this for you because in MVVM there is supposed to be SoC between the `View` and `ViewModel` classes and between the `ViewModel` and `Model` classes. This is especially true with an ArcGIS Runtime app, because we could have data from both online and offline services. So far, we've been using online services to illustrate the basic application design. But, what if you had dozens of layers and you need to perhaps change out the URLs or the paths to local data? We need a simple solution that we can build on. Also, before we continue, we should go ahead and decide the ways we want to store our configuration.

There are many ways to tackle this problem:

* Create a file and store our configuration there
* Create a web service and store our configuration there
* Use both the preceding options

Using the first option is fine, but if your users will need to access the app from multiple devices, this means they would have two or more configurations. If you go with the second option, this means you would need constant Internet access, which goes against the grain when it comes to native apps, because they usually need to work offline at some point. Using the third option seems to be ideal.

If there's Internet access, use the latest configuration on the Web. If you get disconnected, use the latest local configuration. If you make a change while offline, prompt the user when they go back online to upload the latest changes. Once again, these are all design decisions that need to be well thought out.

For this section, we're going to keep it simple and use the first option. We're going to store the URLs for our layers in a simple file and have the ViewModel class call a service, which loads them in the Model class for us from a JSON file. That way, we can change the app and it should work just like it does with hardcoded values.

To accomplish this, we need a way to read JSON. For this exercise, we're going to use a third-party JSON library:

1. Make a new project. In the following example, it's going to be called Chapter2b.

2. Using NuGet, add the MVVM Light libraries.

3. Using NuGet, add the ArcGIS Runtime SDK for .NET v10.2.7.

4. Using NuGet, add NewtonSoft.Json, as shown in the following screenshot:

5. Copy `Model.cs` from the previous project. Be sure to update the namespace in this file to `Chapter2b.Models`.

6. When you use NuGet to add the MVVM Light libraries, it will create the `MainViewModel.cs` and `ViewModelLocator.cs` files for you, so just copy the contents of the previous `MainViewModel.cs` file to the new one. However, the `MainViewModel` property needs to be renamed from `Main` to `MainViewModel`, as shown here:

```
public MainViewModel MainViewModel
{
    get
    {
        return
        ServiceLocator.Current.GetInstance
            <MainViewModel>();
    }
}
```

7. Copy the contents from the `MainWindow.xaml` file to the new one. Update the references at the top of the XAML file. For example, change them from `Chapter2a` to `Chapter2b`.

8. Make sure your `App.xaml` file looks like this:

```
<Application x:Class="Chapter2b.App"
    xmlns="http://schemas.microsoft.com/winfx/2006/xaml/
        presentation"
    xmlns:x="http://schemas.microsoft.com/winfx/2006/xaml"
        StartupUri="MainWindow.xaml"
            Startup="Application_Startup"
    xmlns:d="http://schemas.microsoft.com/expression/blend/
        2008"
    xmlns:vm="clr-namespace:Chapter2b.ViewModels"
        d1p1:Ignorable="d"
    xmlns:d1p1="http://schemas.openxmlformats.org/
        markup-compatibility/2006">
    <Application.Resources>
        <ResourceDictionary>
            <vm:ViewModelLocator x:Key="Locator"
                d:IsDataSource="True"  />
        </ResourceDictionary>
    </Application.Resources>
</Application>
```

9. Add a new folder to the project. Name it `Services`.

10. Add a new generic class. Name it `ConfigService.cs` and make it look like the following code:

```
using System;
using System.Collections.Generic;
using System.Linq;
using System.Text;
using System.Threading.Tasks;
using System.IO;

using Newtonsoft.Json;
using Chapter2b.Models;

namespace Chapter2b.Services
{
    public class ConfigService
    {
        public Model LoadJSON()
        {
            string modelContent =
                File.ReadAllText(@"C:\
                    C:\ArcGISRuntimeBook\JSON\
                        config.json");
            return
            JsonConvert.DeserializeObject<Model>
                (modelContent);
        }
    }
}
```

11. In `MainViewModel.cs`, add a reference to `Chapter2b.Services`, and then change the constructor to the following code:

```
public MainViewModel()
{
    if (IsInDesignMode)
    {
        // Code runs in Blend --> create design time data.
    }
    else
    {
        // Code runs "for real"
        ConfigService config = new ConfigService();
        this.myModel = config.LoadJSON();
```

```
        this.SearchRelayCommand = new
        RelayCommand<int>(Search);
    }
}
```

12. Open the JSON file located at `C:\ArcGISRuntimeBook\JSON\config.json` and note that its contents look like this:

```
{"SearchText":"Lancaster",
    "BasemapLayerUri":"http://services.arcgisonline.com/
        ArcGIS/rest/services/World_Topo_Map/MapServer",
    "USALayerUri":"http://sampleserver6.arcgisonline.com/
        arcgis/rest/services/USA/MapServer"}
```

13. Run the app. Once again, nothing much has changed; but this time, we've used a service to load our data from a configuration file, so now our app can be configured. Refer to the sample project called `Chapter2b` provided with this book for the complete app.

In a production-level app, this design could be greatly improved on by creating a separate configuration tool that users use to configure the app before it is started. Or, you could make this app have all the configuration options built into it.

Displaying a message

We're almost done with our basic app. As you will recall, earlier we commented out the `MessageBox` component. The reason we did that was because a `MessageBox` component is a UI component, which means it shouldn't be in the `ViewModel` class. How do we resolve this? Well, luckily MVVM Light comes to the rescue with the **Messenger** service. The Messenger service does as its name implies: it allows you to send messages in the form of a service pattern. Making a few minor changes to our app allows us to maintain SoC while showing the user a `MessageBox` component. Here are the steps:

1. In the `ViewModelLocator` class, add a `using` statement to `GalaSoft.MvvmLight.Messaging` and `System.Windows`.

2. In the constructor of `ViewModelLocator`, add the following line at the end of the constructor:

```
Messenger.Default.Register<NotificationMessage>(this,
    NotifyUserMethod);
```

3. Add a method to `ViewModelLocator`, as shown in this code:

```
private void NotifyUserMethod(NotificationMessage message)
    {
        MessageBox.Show(message.Notification);
    }
```

4. Back in the `MainViewModel` instance, add a `using` statement to `GalaSoft.`
`MvvLight.Messaging`, and then add the following line to the end of the
`Search` method:

```
Messenger.Default.Send<NotificationMessage>(new
    NotificationMessage(msg));
```

5. You can remove this line:

```
System.Diagnostics.Debug.WriteLine(msg);
```

6. While you're at it, go ahead and remove this line in the `Search` method:

```
var url =
    "http://sampleserver6.arcgisonline.com/arcgis/rest/
        services/USA/MapServer";
```

7. Then, have a look at this line:

```
var findTask = new FindTask(new System.Uri(url));
```

8. Replace the preceding line with this one:

```
var findTask = new FindTask(new System.Uri(this.USAUri));
```

As we've placed our code in a service, and then made it a property on
our `ViewModel` class, there's no sense in having a hardcoded URL in the
`Search` method.

9. Run the app again and click on the **Find** button again. You will see the
`MessageBox` dialog. Once again, the app behaves the same as it has before,
except this time the `MessageBox` dialog has been completely removed from
the `ViewModel` class and is handled by the MVVM Light message service.
When the `Search` method is done with its task, it sends a message to
`NotifyUserMethod`.

These steps have obviously been pretty basic examples, but they illustrate many
of the principles of MVVM. There is a lot more to MVVM and MVVM Light, and
hopefully this has provided you with the impetus to learn more about it, so you can
make your apps meet all of the goals laid out at the beginning of this chapter. You
are encouraged to look into messaging, because it's a powerful technique for sending
messages between different parts of your app. For example, if you have multiple
`ViewModel` classes, messaging is indispensable because you can send messages
between them.

Now that we've written an app using the MVVM pattern, it's time to test this app. The problem is that we really haven't explained many details about many of the code samples we tried. This needs to change so that you have a better understanding of maps, scenes, layers, geocoding, and so on. Because of this, we're going to push testing to *Chapter 12*, *Configuration, Licensing, and Deploying*. By then, you will have a much better understanding of these core concepts and much more.

Summary

In this chapter, you were introduced to the issues involved in implementing an app in .NET using the MVVM pattern. First, you were given the high-level goals of the pattern, next a brief description of the pattern, then the ways MVVM is implemented without a toolkit, and lastly, you were introduced to MVVM Light, which is a toolkit that greatly simplifies issues related to SoC, maintainability, and so on. With this introduction to MVVM Light, you now have the skills and tools to explore the pattern and apply it to your app.

In the next chapter, we're going to do a deeper dive into maps and layers. We will discuss projections and coordinates systems, `MapGrid`, extents, layers, and how to use online and offline content.

3
Maps and Layers

In the previous chapter, you were introduced to ArcGIS Runtime and provided with an example of how to architect your solution without many details on how exactly ArcGIS Runtime works with maps and layers. This chapter's aim is to provide more details on maps and layers and how to apply them to both online and offline content. In short, you will be exposed to the following topics:

- MapView
- Layers
- Using online content
- Using offline content

When you're done with this chapter, you will know how to create and consume online content via ArcGIS Online / Portal for ArcGIS or offline content via a Runtime geodatabase, tile package, map package, and so on.

MapView

In essence, MapView is a container for a map. It resides in the `Esri.ArcGISRuntime.Controls` namespace and is the most important control in ArcGIS Runtime because, without it, you can't view a map. It contains a series of properties, methods, and events just like any other .NET control. However, it's easy to get overwhelmed with all of the options, so let's dissect these options one by one and show some examples of use. Before we look at the details of this control, we first need to understand some mapping concepts.

Projections and coordinate systems

As a mapping developer, you really need to understand projections and coordinate systems. They have a huge impact not only on how data is displayed, but also on performance, usability, and other factors. Projections attempt to portray the surface of the earth on a flat surface. Coordinate systems use a three-dimensional sphere to define a location on the earth's surface.

There are many kinds of projections you can use and it really depends on the location in which your users will be working with your app. If you want a general-purpose method to display your data, latitude/longitude is an excellent choice. However, it has its drawbacks as all projections/coordinate systems have, because if you draw a circle on the map at the equator and then near the poles, the same exact circle will look very different. This is because of the nature of this projections/coordinates system: they distort. In fact, the circle will go from being a circle near the equator to an elliptical shape near the poles:

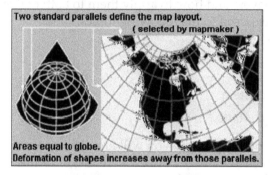

Distortion caused due to the nature of the projections

Choosing the right projections/coordinate system is an important choice because it will affect the following properties of the map:

- Area
- Shape
- Direction
- Bearing
- Distance
- Scale

No matter what projection you choose, at least one of these properties will be affected at the expense of the other properties. For example, if you choose any of the Mercator projections, you'll note that circles become larger if you move from the equator to the poles:

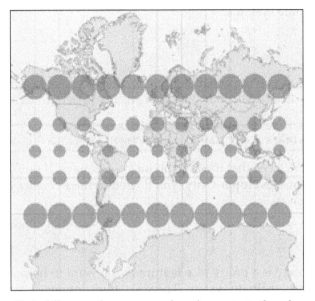

Circle difference when you move from the equator to the poles

For more information about projections and coordinate systems, consult a reputable site on the Web or a book.

When you add a layer to the map, the spatial reference of `MapView` is set by default by the first layer in `MapView`. Because of this, you really don't have to do anything if you're OK with the first layer's projection. Most of the time, the first layer you add will be a basemap layer. (Don't confuse the word basemap with map; it's a layer.) A basemap layer is a nice-looking background layer that allows users to orient themselves in the world. A typical basemap is of the streets, buildings, and other important background information. As you are setting the spatial reference by adding the layer, you can't actually set the spatial reference on `MapView`; you can only get it. You can set the spatial reference of the map object, however.

A spatial reference is a map projection or coordinate system. You can get the spatial reference represented as a **well-known ID (WKID)**, such as 4326, or **well-known text (WKT)**, such as GCS_WGS_1984. The number 4326 just happens to be the **latitude/longitude coordinate system (Lat/Lon)**. Lat/Lon uses angular units for the y and x fields, respectively, and meters for the z (elevation) field. You can find a complete list of coordinate systems at `https://developers.arcgis.com/net/desktop/guide/geographic-coordinate-systems.htm`.

For a complete list of projections, navigate to: `https://developers.arcgis.com/net/desktop/guide/projected-coordinate-systems.htm`.

Coordinate systems come in meters, feet, and other units for the x, y, and z fields, which will be very important later when you learn how to add objects to the map using these coordinate systems.

Esri logo

One of the first things you're probably going to want to do is to remove the Esri logo from `MapView`. In your XAML code, just set `IsLogoVisible` to `false` like this:

```
<esri:MapView x:Name="MyMapView" Grid.Row="0"
  IsLogoVisible="False" LayerLoaded="MyMapView_LayerLoaded">
```

Run the app and you'll see that the logo no longer displays. If you also want to remove **Licensed for Developer Only**, you'll need to read *Chapter 12, Configuration, Licensing, and Deploying*.

MapGrid

The MapGrid property is a pretty nice feature if you want to help your users find a location based on a coordinate system. There are four options:

- **Latitude/Longitude** (`LatLonMapGrid`)
- **Military Grid Reference System** (`MgrsMapGrid`)
- **U.S. National Grid** (`UsngMapGrid`)
- **Universal Transverse Mercator Grid** (`UtmMapGrid`)

To create a grid, just add some XAML code like this, after the `</esri:Map>` tag and before `</esri:MapView>`:

```
<esri:MapView.MapGrid>
    esri:LatLonMapGrid/>
</esri:MapView.MapGrid>
```

The map will look something like this:

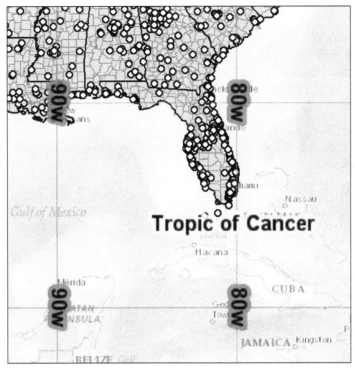

MapGrid

You can change whether to use decimal degrees or `DegreesMinutesSeconds` by using `LabelStyle`. If you want to give your user the ability to turn this option on or off, just create a property on ViewModel and bind to it via `IsVisible`.

Scale

Another option to constrain the map is to set the minimum and maximum scale so that users can only zoom in or out so far. This can be important for novice users because they can zoom in too far and not see any detail and not realize that they just need to zoom out with their mouse or a gesture. The two properties are `MaxScale` and `MinScale`. You can use them alone or together. For example, you may want to restrict the user from zooming in beyond a scale of 1:1,000.

Overlays

Giving your users the ability to click on an object on the map is a nice feature that your users will enjoy when they interact with the map. Otherwise known as MapTips, overlays define what happens when your user clicks on an object using XAML or in code. For example, the following XAML code sets the overlay for a single layer that is using a field named MYFIELD from a layer named FeatureLayer:

```xaml
<esri:MapView.Overlays>
    <esri:OverlayItemsControl x:Name="overlayItemsControl">
        <Border
            CornerRadius="10"
            BorderBrush="Black"
            Margin="0,0,25,25" Visibility="Hidden"
            BorderThickness="2" Background="#995C90B2" >

            <StackPanel Orientation="Vertical"
                Margin="5,10,18,15">
                <StackPanel Orientation="Horizontal">
                    <TextBlock Text="" FontWeight="Normal"
                    Foreground="Black"/>
                </StackPanel>
                <Line X1="0" X2="200" Y1="0"  Y2="0"
                    Stroke="White"
                StrokeThickness="2" />
                <StackPanel Orientation="Horizontal">
                    <TextBlock Text="Name: " FontWeight="Normal"
                    Foreground="Black"/>
                    <TextBlock Text="{Binding [MYFIELD]}"
                        FontWeight="Normal"
                    Foreground="Black" />
                </StackPanel>
            </StackPanel>
        </Border>
    </esri:OverlayItemsControl>
</esri:MapView.Overlays>
```

The FeatureLayers are a kind of layer that contains geometry such as points, lines, polygons, and attributes. With an overlay, you can click on a point and see information about that point as shown here:

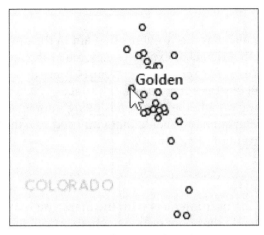

Displays the information about the point

Editing

In order to edit layers that allow editing, the MapView container contains an Editor class. Once you have a reference to MapView, the Editor class is available to add, update, and delete objects on the map, such as points, lines, arrows, triangles, and polygons. With this single class, you can edit both online and offline layers, which makes editing pretty easy to accomplish. Editing not only includes the ability to manipulate geometry, it also includes the ability to edit attributes (fields) of the data too. If the data resides in an online service, the edits you make will be immediately reflected online. If you allow your users to edit offline data, the edits will be stored in a local store until you synchronize them to an online service or copy them to an enterprise geodatabase for other uses.

WrapAround

WrapAround is a nice property if you want to give your users the ability to pan east or west around the globe. When they pan the map, they will automatically go around the International Date Line. Just set this property to true to activate it.

LocationDisplay

With the LocationDisplay property on the MapView container, you can get your current location via a location provider. A location provider can be the GPS in your device, an attached GPS, or a custom location provider, such as a random location simulator.

Events

The `MapView` container also has many events that aid in the development of providing your app with a rich user experience. Some of the more interesting and useful events are `Initialized`, `MapViewDoubleTapped`, `MouseDoubleClick`, `MouseDown`, `MouseMove`, `StylusButtonDown`, `StylusDown`, `TouchDown`, `TouchMove`, and `TouchUp`. Depending on whether the app is being interacted with via mouse, stylus, or touch, will determine the events that you need to handle and the requirements you are trying to implement.

Map interactions

When it comes to how the user interacts with the map, you have many options you can control. You can enable the zoom options, whether double-tapping is enabled, whether a zoom box is enabled, whether rotation and panning are allowed, and so on. To set this in XAML, add the following example code after you've defined the `MapView` container and before the `Map` XAML line:

```
<esri:MapView.InteractionOptions>
    <esri:InteractionOptions>
        <esri:InteractionOptions.ZoomOptions>
            <esri:ZoomOptions IsEnabled="True"
                IsDoubleTappedEnabled="False"
                IsZoomBoxEnabled="True"
                IsMouseWheelEnabled="True"
                IsPinchEnabled="True"
                IsTwoFingerTapEnabled="True"
            IsKeyboardEnabled="False" />
        </esri:InteractionOptions.ZoomOptions>
        <esri:InteractionOptions.RotationOptions >
            <esri:RotationOptions IsEnabled="False" />
        </esri:InteractionOptions.RotationOptions>
        <esri:InteractionOptions.PanOptions>
            <esri:PanOptions IsEnabled="True"
                IsDragEnabled="True"
                IsFlickEnabled="True"
            IsKeyboardEnabled="True" />
        </esri:InteractionOptions.PanOptions>
    </esri:InteractionOptions>
</esri:MapView.InteractionOptions>
```

Map

Now that you've been introduced to the `MapView` container, it's time to get to the main attraction: the map. The map is really a container for layers, but in order to do just about anything, you will need to access the map directly or indirectly via the `MapView` container so that you can do things with layers. The map has three properties that you can interact with in XAML: `InitialViewpoint`, `Layers`, and `SpatialReference`. Most of the other important methods, properties, events, and so on are controlled with code.

Extent

Every time you zoom in and out, panning whether with a mouse or your finger, you are changing the extent of the map you are viewing. The extent is set with coordinates, hence it's important for you to understand your coordinate system. The extent is rectangular with a minimum and maximum x and minimum and maximum y value. You can set the initial extent of the map in XAML like this:

```
<esri:MapView x:Name="MyMapView">
    <esri:Map x:Name="MyMap">
        <esri:Map.InitialViewpoint>
            <esri:Map.InitialViewpoint >
                <esri:ViewpointExtent XMin="12.34" YMin="41.85"
                XMax="12.55" YMax="41.95"
                SpatialReferenceID="4326" />
            </esri:Map.InitialViewpoint>
        </esri:Map.InitialViewpoint>
    </esri:Map>
</esri:MapView>
```

You're now looking at Rome, Italy. This will work no matter the projection of the first layer in the map because the spatial reference was also set using `SpatialReferenceID`. If your map shows up in an incorrect location, the first thing to check is the extent coordinates and the spatial reference.

You can also just use `ViewpointCenter` to set the x and y fields and the scale too. This property will place the map at the exact x and y coordinates and the scale you provide.

These are the only two parameters that you can use to set the extent using XAML. All other parameters have to be set with code, which would reside in the `ViewModel` classes. Setting the extent is handled with the following methods: `SetView`, `SetViewAsync`, `ZoomAsync`, and `ZoomToScaleAsync`. The extent can be set with `MapPoint` and `Viewpoint` with geometry such as a bounding rectangle, or scale. Here is an example:

```
MapPoint point =  new Esri.ArcGISRuntime.Geometry.MapPoint(12.5,
    41.9, new SpatialReference(4326));

mapView.SetViewAsync(point, 100000);
```

This code has the same effect as the previous code except that it zooms out to a scale of 1:100,000 around Rome.

Layers

Layers typically come in three flavors when you add them to the map:

- Basemap layers
- Operational layers
- Dynamic layers

The basemap layers are typically static layers that usually show a collection of layers, such as streets, imagery, oceans, states, and country boundaries in a single layer. They are generally speaking view only and have been designed to be fast. The operational layers are typically layers that users view, query, and edit. These kinds of layers serve the same purpose as transactional tables in a relational database. The operational layers are viewed on top of the basemap layers. They can also be edited and persisted to a geodatabase. Lastly, the dynamic layers are the layers that are updated frequently, even in real time. This includes temporary graphics on the map, such as tracking aircraft in real time or the current position of the user as they drive down a street. Once the app closes or the user stops tracking themselves, these kinds of layers are not persisted into the geodatabase or any other persistence mechanism.

The `MapView` and `Map` containers don't do you much good unless they have layers in them. ArcGIS Runtime supports many types of layers. In this section, we will explore them so that you can get a feel for all of the options. All layers are located in `Esri.ArcGISRuntime.Layers`. You've already seen a couple of layers in the previous chapter when you added XAML like this:

```
<esri:ArcGISTiledMapServiceLayer ID="Basemap"
    ServiceUri="{Binding Source={StaticResource VM},
Path=BasemapUri}"/>
<esri:ArcGISDynamicMapServiceLayer ID="USA"
```

```
       ServiceUri="{Binding Source={StaticResource VM},
   Path=USAUri}"/>
```

But what exactly is a tiled map service or a dynamic map service? What kinds of layers are supported? Refer to the diagram shown here:

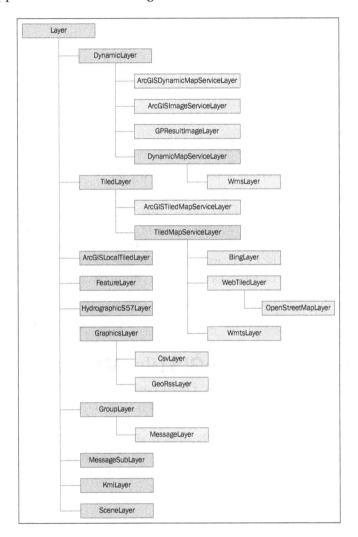

As can be seen in the object hierarchy, `Layer` is the base class. All subclasses inherit all of its properties and methods. In the second tier, there are the following classes:

- `DynamicLayer`
- `TiledLayer`
- `ArcGISLocalTiledLayer`

- FeatureLayer
- HydrographicS57Layer
- GraphicsLayer
- GroupLayer
- MessageSubLayer
- KmlLayer
- SceneLayer

Let's start with some of the ones we've already been exposed to, and then build our understanding from there forward.

Dynamic map service – ArcGISDynamicMapServiceLayer

A dynamic map service is a map with one or more layers, which users can interact with just as they do from ArcGIS Desktop. In other words, users are able to view and interact with the geometry and fields of the layers. A user can search, change symbology if enabled, identify, and so on. Every time the user zooms or pans, an image is generated on the fly. This can be an expensive process because this step involves taking a snapshot of the extent with the geometry and sending it to the client.

To access these services, all you have to do is specify the REST endpoint from either in the cloud via AGOL or from your own implementation of the ArcGIS Server.

Tiled map service – ArcGISTiledMapServiceLayer

You can think of a tiled map service as a high-performance approach to taking many geometry-based (vector) and/or imagery-based (raster) layers and converting them to tiled images. When you look at one of these layers across the entire earth, you are actually looking at a few images, which are actually snapshots. Basically, the earth is being divided up into a series of tiled images. So, instead of seeing the original geometry, which would be too slow to visualize, you are seeing pictures broken up at different scales.

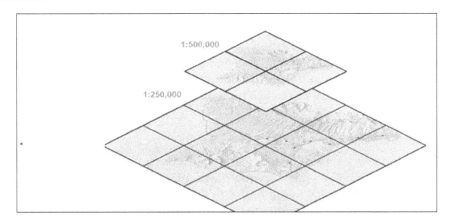

Many of these layers are called basemaps because they contain a set of layers that have been merged to form a picture that is fast and can move quickly across the Internet due to the fact that the images have been compressed in some cases.

Creating these tiles is done with ArcGIS Desktop, and then published to AGOL or ArcGIS Server. Although the process to generate these tiles can be slow, sometimes very slow, the end result is that users of these kinds of layers get very fast rendering times because they are just looking at pictures (tiles). As we are only viewing tiles, you can't do anything else with the tiles, such as querying, identifying, and the other things you can do with dynamic map services and feature services.

As you've seen many times now, all you have to do is specify the URI and name and you now have a basemap layer. There are many basemaps available, which has lessened the need for most organizations to develop their own.

Navigate to the following link to see a complete list:

```
http://www.arcgis.com/home/group.html?owner=esri&title=Tiled%20
Basemaps
```

Click on **Details** under the snapshot of the basemap, and then find the REST endpoint in the description or **Map Contents** section. You can use as many of these maps as you want, but always remember that the more you add, the slower the performance, so choose the best one for your app.

Local tiled layer – ArcGISLocalTiledLayer

A local tiled layer is similar in concept to the `ArcGISTiledMapServiceLayer`, except that it resides on a local disk for disconnected viewing. The tiles are cached to a disk using ArcGIS Desktop into a single `.tpk` file and copied to the device for offline use. As such, it is very fast because it doesn't rely on an Internet connection.

Feature service – FeatureLayer

The feature service is a service in some ways like a dynamic map service except that no image is generated when it is drawn. What actually happens is that the geometry and attributes are returned across the Web in the form of JSON to the client, which users can view, query, change symbology, and edit. The editing capability is the primary reason to use the feature service. This is an important type of service because it allows for both connected and disconnected editing on a device. In fact, with this service, you can download data, go disconnected, perform edits, and then synchronize your edits when you go back online. Or, if you're always connected, you can just edit this service directly at all times.

To create this kind of layer, you must create it using ArcGIS Desktop. It is, by default, not enabled when a service is created. Once the service is created, you can access it by using the following code:

```
<esri:FeatureLayer ID="Incidents">
    <esri:ServiceFeatureTable
        ServiceUri="http://sampleserver6.arcgisonline.com/arcgis/
            rest/services/SF311/FeatureServer/0" />
</esri:FeatureLayer>
```

Note that in the REST endpoint, it denotes this service as `FeatureServer/0`. This indicates that it's a feature service and we're accessing the zero layer, which is the first layer in this service. Also, note that we define the layer using `FeatureLayer`, and then inside the `FeatureLayer` class, there is a `ServiceFeatureTable` property. Using `ServiceFeatureTable` indicates that this service is coming from an online source. Alternatively, you can use a `FeatureLayer` class to access an offline data source coming from an ArcGIS Runtime geodatabase via `ArcGISFeatureTable`. The first thing you will need to do is open the Runtime geodatabase with the following code:

```
var gdb = await Geodatabase.OpenAsync(gdbPath);
```

Once you have an access to the geodatabase, you can find and add layers to the map with the following code:

```
// loop thru all tables in the geodatabase
foreach (var table in gdb.FeatureTables.ToList())
{
```

```
    // create a new feature layer to display each table
    var lyr = new FeatureLayer
    {
        ID = table.Name + "Offline",
        DisplayName = table.Name,
        FeatureTable = table
    };

    // add each layer to the map
    MyMapView.Map.Layers.Add(lyr);
}
```

If you navigate to `http://sampleserver6.arcgisonline.com/arcgis/ rest/services`, you will find many feature services. Any service listed with `(FeatureServer)` beside its name is `FeatureService`:

- <u>911CallsHotspot</u> (GPServer)
- <u>911CallsHotspot</u> (MapServer)
- <u>Census</u> (MapServer)
- <u>CharlotteLAS</u> (ImageServer)
- <u>CommercialDamageAssessment</u> (FeatureServer)
- <u>CommercialDamageAssessment</u> (MapServer)
- <u>CommunityAddressing</u> (FeatureServer)
- <u>CommunityAddressing</u> (MapServer)
- <u>DamageAssessment</u> (FeatureServer)
- <u>DamageAssessment</u> (MapServer)
- <u>DamageAssessmentStatePlane</u> (FeatureServer)
- <u>DamageAssessmentStatePlane</u> (MapServer)
- <u>EmergencyFacilities</u> (FeatureServer)

Later on in this chapter, we will explore how to open a Runtime geodatabase without using a feature service. Also, in *Chapter 9, Editing Features*, we will explore the editing pattern in much more detail.

Image services – ArcGISImageServiceLayer

The image services are accessed via the `ArcGISImageServiceLayer`. These kinds of layers can come from both online and offline local sources in the form of aerial photography, for example. To add one of these kinds of services from XAML, use this code:

```
<esri:ArcGISImageServiceLayer ID="Imagery of Toronto"
    ServiceUri="http://sampleserver6.arcgisonline.com/arcgis/rest/
        services/Toronto/ImageServer">

</esri:ArcGISImageServiceLayer>
```

The following site has some image services that you can use: `https://sampleserver3.arcgisonline.com/ArcGIS/rest/services/World`.

Graphics layers – GraphicsLayer

A graphics layer is used for showing dynamic content or any content that is temporary in nature. For example, you may want to give your user the ability to mark up a map or show the real-time position of a fast-moving drone on the map. You can add as many graphic layers as you need, but typically, you'll find that you need three or less. To add a graphics layer via code, you'd enter something like the following:

```
var graphicsLayer = new Esri.ArcGISRuntime.Layers.GraphicsLayer();
graphicsLayer.ID = "MyGraphicsLayer";

// add the graphics layer to the map
MyMapView.Map.Layers.Add(graphicsLayer);
```

However, this only creates the layer; it doesn't actually add anything to the layer. To accomplish this, you will need to input the following code:

```
// create a symbol (large cyan circle) for displaying a point
var markerSym = new
    Esri.ArcGISRuntime.Symbology.SimpleMarkerSymbol();
markerSym.Color = Colors.Cyan;

markerSym.Size = 18;
markerSym.Style =
    Esri.ArcGISRuntime.Symbology.SimpleMarkerStyle.Circle;

// create a marker graphic for the center of the map's extent
var pointGraphic = new
    Esri.ArcGISRuntime.Layers.Graphic(MyMapView.Extent.
        GetCenter(), markerSym);

// add the graphic to the graphics layer for display on the map
graphicsLayer.Graphics.Add(pointGraphic);
```

This code creates a symbol with a certain color, size, and style, and then creates a graphic from a point using the center of the map, and finally adds the resulting graphic to the graphics layer. More details on symbology will be covered in *Chapter 5, Geometry and Symbology*.

CSV layer – CsvLayer

A CSV layer is a layer from a CSV file, which contains text data that is delimited by some characters for each column. Each line of the file contains a row. The `CsvLayer` class is a type of `GraphicsLayer`.

Bing layers – BingLayer

A Bing layer is a service provided by Microsoft's Bing tiled map services. The key difference between this service and other services that we've discussed is that it's accessed via SOAP as opposed to REST. Also, you must acquire a key using a Bing account from `https://www.bingmapsportal.com/`.

Once you've acquired this key, you can add a layer using the following code:

```
var bingMapLayer = new Esri.ArcGISRuntime.Layers.BingLayer();
bingMapLayer.Key = "gqw%2ss@2";
bingMapLayer.MapStyle =
    Esri.ArcGISRuntime.Layers.BingLayer.LayerType.
        AerialWithLabels;
```

The Keyhole Markup Language layers – KmlLayer

Keyhole Markup Language (**KML**) is an XML notation for representing geographic data in 2D and 3D in Google Maps and Google Earth. Since its inception in 2004, it has become very popular with both GIS and non-GIS professionals as a means to make layers. To add a layer via code to the map is just as easy as the previous examples:

```
// create a KML layer from an online file
var sourceUri = new
    Uri("http://earthquake.usgs.gov/earthquakes/feed/v1.0/summary/
        1.0_week_age_link.kml");
var kmlLayer = new KmlLayer(sourceUri);

// Add the kml layer to the map
MyMapView.Map.Layers.Add(kmlLayer);
```

Group layers – GroupLayer

A group layer is a mechanism to logically group your layers. For example, you can group all of our operational layers (FeatureLayers) into a group so that when they show up in a legend they are under one branch (folder) in the legend. The group layers don't store data themselves, they just contain other kinds of layers. You can create as many group layers as you need. Here is an example of creating a group layer in XAML:

```
<esri:GroupLayer DisplayName="Basemap Group">
    <esri:ArcGISTiledMapServiceLayer
        DisplayName="Imagery" IsVisible="False"
        ServiceUri="http://services.arcgisonline.com/ArcGIS/rest/
            services/World_Imagery/MapServer" />
    <esri:ArcGISTiledMapServiceLayer DisplayName="Street"
        ServiceUri="http://services.arcgisonline.com/ArcGIS/rest/
            services/World_Street_Map/MapServer" />
</esri:GroupLayer>
```

Message layer – MessageLayer

A message layer displays military symbols using graphics that conform to the MIL-STD-2525C or APP6B standards by having a message using a **Symbol ID Code (SIC)** sent to this layer type.

Message sublayer – MessageSubLayer

A message sublayer is a sublayer of a message layer and it's where messages are sent for display on the map.

Scene layer – SceneLayer

A scene layer is a layer for rendering 3D content in a scene/scene view. This kind of layer can't be viewed in a MapView/Map container. With this kind of layer, 3D content such as buildings, trees, and transmission towers, and generic 3D objects such as spheres, pyramids, and cones can also be viewed. This kind of layer will be discussed in the next chapter.

Web map service layer – WMSLayer

The web map service layer allow you to add **Open Geospatial Consortium (OGC)** services to a map. Typically, this layer is a basemap layer or an operational layer in the form of a dynamic map service.

OpenStreetMap – OpenStreetMapLayer

The OpenStreetMap editor is a crowd source basemap based on tiles, which you can utilize via the OpenStreetMapLayer.

Using online content

Now that you've been exposed to some concepts, let's put this into action by building an app that allows you to explore MapView, Map, and some of the layers in more detail than we've done in previous chapters. While we're at it, let's add some functionalities to the app, such as a button that zooms to a certain location, setting Viewpoint, removing the Esri logo, and setting the interaction options:

1. Make a copy of the last project from *Chapter 2*, *The MVVM Pattern* (Chapter2b). Make sure to resolve any errors and make sure it will build and run as it did before.

   In order to make some of the functionalities, such as accessing the Editor class, we will need to make sure our ViewModel class has access to the MapView container; otherwise, our app won't work and we wouldn't satisfy SoC. To achieve this, we will customize our MapView container to use a custom behavior. Behaviors are not a part of MVVM Light, but that's OK because we can go beyond MVVM Light while at the same time using more advanced features of WPF while achieving all of the goals of MVVM.

2. Add a reference to System.Windows.Interactivity.

3. Create a new folder in your project and name it Behaviors. Add a new class and name it MapViewBehavior.cs.

4. Enter the following code in MapViewBehavior.cs:

```
using Esri.ArcGISRuntime.Controls;
using GalaSoft.MvvmLight.Messaging;
using System;
using System.Collections.Generic;
using System.Linq;
using System.Text;
using System.Threading.Tasks;
using System.Windows;
using System.Windows.Interactivity;

namespace Chapter3.Behaviors
{
    public class MapViewBehavior : Behavior<MapView>
    {
```

```
protected override void OnAttached()
{
    base.OnAttached();
    AssociatedObject.Loaded +=
        AssociatedObject_Loaded;
}
void AssociatedObject_Loaded(object sender,
    RoutedEventArgs e)
{

    MapView mapView = sender as MapView;
    Messenger.Default.Send<MapView>(mapView);
}
        }
    }
```

As we discussed in the previous chapter, a behavior is mechanism to change how an element behaves. In this case, we're just making the `MapView` container send a reference to anything that is listening to the MVVM Light messenger. In the constructor, `OnAttached` is called as soon as the `MapView` class is initiated. Then, we wire up an event handler to handle when the `MapView` class is loaded. In the event handler, we get a reference to the `MapView` class and send it to `Messenger`. Anything listening will get the `MapView` class. We are now going to make our `ViewModel` class receive this `MapView` class:

1. In `MainViewModel.cs`, add a reference to `Esri.ArcGISRuntime.Controls`.

2. Add a private member to the calls like this:

   ```
   private MapView mapView = null;
   ```

3. Add the following lines after the `this.SearchRelayCommand` as shown here:

   ```
   public MainViewModel()
   {
       if (IsInDesignMode)
       {
           // Code runs in Blend --> create design time data.
       }
       else
       {
           // Code runs "for real"
           ConfigService config = new ConfigService();
           this.myModel = config.LoadJSON();
   ```

```
        this.SearchRelayCommand = new
            RelayCommand<int>(Search);

        // added
        Messenger.Default.Register<Esri.ArcGISRuntime.
            Controls.MapView>(this, (mapView) =>
        {
            this.mapView = mapView;

        });

    }
}
```

4. Open `MainWindow.xaml` and add the following two references:

```
xmlns:behave="clr-namespace:Chapter3.Behaviors"
xmlns:i="clr-
    namespace:System.Windows.Interactivity;assembly=System.
        Windows.Interactivity"
```

5. Right after the `MapView` class is defined with XAML, add the following lines:

```
<i:Interaction.Behaviors>
    <behave:MapViewBehavior/>
</i:Interaction.Behaviors>
```

6. Run the app. If you put a break on the line where the `MapView` class is set in the constructor (`this.mapView = mapView`), you will see that it captures a reference to the `MapView` class. There you have it! Your `ViewModel` class is now set to do just about anything because it now has access to the `MapView` class, which gives it access to everything of relevance for building a complete app.

7. Add a button on top of the map like this:

```
<Button Content="Zoom To Italy" Width="80" Height="35"
    HorizontalAlignment="Left"
    Command="{Binding ZoomRelayCommand}">
</Button>
```

8. Add a new relay command to `MainViewModel` like this:

```
public RelayCommand ZoomRelayCommand { get; private set; }
```

9. Modify the constructor of `MainViewModel` to instantiate the `RelayCommand` class like this:

```
this.ZoomRelayCommand = new RelayCommand(Zoom);
```

10. Add the following method to the `ViewModel` class:

```
public async void Zoom()
{
    MapPoint point = new
        Esri.ArcGISRuntime.Geometry.MapPoint(12.5,
        41.9, new SpatialReference(4326));
    await mapView.SetViewAsync(point, 100000);
}
```

11. Add the following method to the `ViewModel` class:

```
public void SetInitialExtent()
{
    // Set the initial extent.
    Esri.ArcGISRuntime.Geometry.SpatialReference sr =
        Esri.ArcGISRuntime.Geometry.SpatialReferences.
            Wgs84;
    Esri.ArcGISRuntime.Geometry.Envelope envelope = new
        Esri.ArcGISRuntime.Geometry.Envelope(10, 40, 14,
            43, sr);
    this.mapView.Map.InitialViewpoint = new
        Esri.ArcGISRuntime.Controls.Viewpoint(envelope);
}
```

12. Add the following line at the end of the anonymous method of the `Messenger` class:

```
this.SetInitialExtent();
```

13. Run the app. You will note that you are zoomed in over Italy because `InitialViewpoint` is set to an extent over Italy. Next, click on **Zoom To Italy**. The `Viewpoint` button is set using an envelope, which is a rectangle that has been set with latitude and longitude coordinates because we specified a spatial reference of WGS84.

While it's true that we could have just passed in the `MapView` class similarly to how we did in the previous chapter, that doesn't satisfy other situations we'll encounter. For example, if we had another `ViewModel` class, how would we pass the `MapView` class to it? What if we didn't have a button at all? What if we need to perform some logic on what layers to load without using XAML at all? What if we wanted to set up MapTips dynamically to multiple layers in the `ViewModel` class without knowing the layers that we'd load up? This is a powerful technique that will now allow us to write our apps without having to be concerned with getting the `MapView` class in our `ViewModel` class. It's just there and we can now focus on the requirements.

One last thing before we move on. The astute observer will note that we just placed a reference to `Esri.ArcGISRuntime.Controls` in our `ViewModel` class. This breaks SoC, but it can easily be resolved by passing the `MapView` class directly to a service without adding the `using` statement to the `ViewModel` class. This was done to illustrate the concept because we only have one method we're calling on, called `Zoom`. In a production-level app, we'd add a service that handled all of our navigation operations. Better yet, we'd place this in its own `ViewModel` class that used this service:

1. Let's configure the map, but let's add it in the `ViewModel` class instead of XAML, so we can see that we can now configure our map using this approach. In `MainViewModel.cs`, update `Messenger` as shown in this code:

```
Messenger.Default.Register<Esri.ArcGISRuntime.Controls.
    MapView>(this, (mapView) =>
{
    this.mapView = mapView;

    this.mapView.MapGrid = new
    LatLonMapGrid(LatLonMapGridLabelStyle.
        DegreesMinutesSeconds);

    this.mapView.InteractionOptions.PanOptions.IsEnabled =
        true;
    this.mapView.InteractionOptions.PanOptions.
        IsDragEnabled = true;
    this.mapView.InteractionOptions.PanOptions.
        IsFlickEnabled = true;
    this.mapView.InteractionOptions.PanOptions.
        IsKeyboardEnabled =
        true;
    this.mapView.InteractionOptions.ZoomOptions.IsEnabled =
        true;
    this.mapView.InteractionOptions.ZoomOptions.
        IsDoubleTappedEnabled
        = false;
    this.mapView.InteractionOptions.ZoomOptions.
        IsMouseWheelEnabled =
        true;
    this.mapView.InteractionOptions.ZoomOptions.
        IsPinchEnabled = true;
    this.mapView.InteractionOptions.ZoomOptions.
        IsTwoFingerTapEnabled
        = true;
    this.mapView.InteractionOptions.ZoomOptions.
        IsKeyboardEnabled =
        true;
});
```

This is pretty nice. We've configured how users can interact with the map but we've placed this code in the `ViewModel` class instead of XAML.

2. Add a reference to `Esri.ArcGISRuntime.Layers` and `System`, and then at the end of this same anonymous method, add the following layers using the following code:

```
Uri uriBasemap = new Uri(this.BasemapUri);
ArcGISTiledMapServiceLayer basemapLayer = new
    ArcGISTiledMapServiceLayer(uriBasemap);

Uri uriUSA = new Uri(this.USAUri);
ArcGISDynamicMapServiceLayer dynamicMapLayer = new
    ArcGISDynamicMapServiceLayer(uriUSA);

this.mapView.Map.Layers.Add(basemapLayer);
this.mapView.Map.Layers.Add(dynamicMapLayer);
```

Just like with the XAML code we've been using before, we've added the layers.

3. Remove the XAML code from `MainWindow.xaml` between `<esriMap>` and `</esri:Map>`, and then run the app. You should see the same layers you've been seeing but this time the layers are coming from the `ViewModel` class via `Messenger`. See the sample app named `Chapter3` provided with this book if you need any help.

At this point, we have a good architecture to build on. We now know how to use MVVM Light to the point that we could extend this to use a service factory so that we completely satisfy SoC. We could even make this service factory load the layers by simply using a `foreach` statement to load them all without even knowing how many layers we have in the configuration file.

The other important point to understand about layers is that each layer is drawn in the order in which it is added. The basemap was added first, so it will be drawn first. The USA layer was added second, so it will be drawn after the basemap.

Using offline content

So far, we've been working with online services, which are great, but ArcGIS Runtime allows you to use offline content too. In this section, we're going to explore three ways of using offline content: tile package, map package, and Runtime geodatabase. If you have ArcGIS Desktop, feel free to use your own data. However, in the following examples, we're going to use data provided with this book for the city of San Francisco.

Local tiled layers

A local tiled package is similar in concept to a tiled service from an online source except that it is saved into a file, which can be side loaded on a device. Refer to the following ArcGIS site for help and more details about creating a tile package: `http://desktop.arcgis.com/en/desktop/latest/map/working-with-arcmap/how-to-create-a-tile-package.htm`.

Let's create a new app based on San Francisco that allows us to visualize this great city using the MVVM pattern in case we need to have offline access:

1. Create a new project and copy the same code over from the previous project. Make sure it contains all of the same classes.

2. Using NuGet, install MVVM Light and `Json.NET`.

3. Update all the namespaces and `using` statements.

4. Remove the XAML code related to the last project, including the `Grid`, `TextBlock` tags, `Find` button, and `Zoom To Italy` button.

5. We will need the configuration file, `C:\ArcGISRuntimeBook\JSON\config_sf.json`, which has the following contents:

   ° **Tile package**: `C:\\ArcGISRuntimeBook\\Data\\SanFrancisco.tpk`

   ° **Map package**: `C:\\ArcGISRuntimeBook\\Data\\MyRestaurants.mpk`

   ° **GDB**: `C:\\ArcGISRuntimeBook\\Data\\sf_parking_meters.geodatabase`

6. Update your `config` service to this location of the `.json` file.

7. Remove the `Search`, `Zoom`, and `SetIntialExtent` methods in the `ViewModel` class. Also, remove `SearchRelayCommand` and `ZoomRelayCommand`. Remove the property named `GridDataResults`.

8. Update the `Model` class to have these three properties: `TilePackage` as a string, `MapPackage` as a string, and `gdb` as a string. Make sure to update the `ViewModel` class with the new property names. Remove the `Search` method, and all the code for adding the layers. Remove all XAML in the `Map` class.

9. In the `ViewModel` class, add the following code in the anonymous method of the `Messenger` class:

```
Messenger.Default.Register<Esri.ArcGISRuntime.Controls.
    MapView>(this, (mapView) =>
{
    this.mapView = mapView;
    ArcGISLocalTiledLayer localTiledLayer = new
```

```
              ArcGISLocalTiledLayer(this.TilePackage);
          localTiledLayer.ID = "SF Basemap";
          localTiledLayer.InitializeAsync();

          this.mapView.Map.Layers.Add(localTiledLayer);
      });
```

10. Run the app and you will see the tile package of San Francisco:

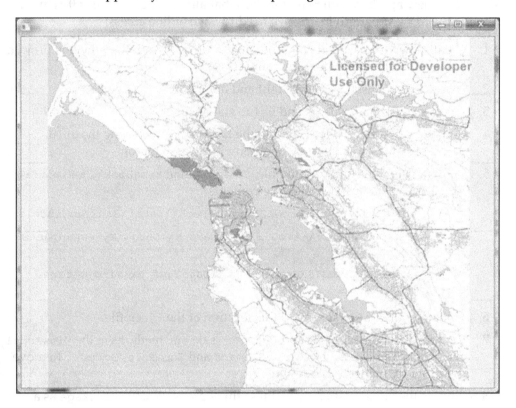

As you can see in this screenshot, our map contains the tile package that covers the bay area. We are no longer using a web service of the world. We had to instantiate the layer using the path to the tile package, initialize it, and then add it to the map. All of this occurred in our `ViewModel` class.

Map package

A map package is simply a compact representation of an ArcGIS Desktop map document. With ArcGIS Runtime, it's possible to view and query this data in the map. This is accomplished by instantiating a local server that allows you to interact with the map as if you were interacting with a map service from ArcGIS Server. Did you catch that? A local server is like a mini version of ArcGIS Server running on your desktop machine. With this local server, you are the administrator and consumer of the data. You have full control over how the local server is configured, including the service life cycle. In fact, it's up to you to start and stop the local server just like you would run ArcGIS Server.

To create a map package via ArcGIS Desktop, navigate here:

```
http://desktop.arcgis.com/en/desktop/latest/map/working-with-arcmap/
creating-a-map-package.htm
```

We will once again use some sample data and make use of a map package:

1. Add a `using` statement in your `ViewModel` class to `Esri.ArcGISRuntime.LocalServices`.

2. Add the following method to the `ViewModel` class:

```
public async void CreateLocalServiceAndDynamicLayer()
{
    LocalMapService localMapService = new
        LocalMapService(this.MapPackage);
    await localMapService.StartAsync();

    ArcGISDynamicMapServiceLayer
        arcGISDynamicMapServiceLayer = new
    ArcGISDynamicMapServiceLayer()
    {
        ID = "Restaurants",
        ServiceUri = localMapService.UrlMapService,
    };


    this.mapView.Map.Layers.Add
        (arcGISDynamicMapServiceLayer);
}
```

3. Call this method at the end of the `Messenger` class just like all of the other initialization code we've added:

```
this.CreateLocalServiceAndDynamicLayer();
```

4. Run the app. You will see a message box that lets you know that this isn't licensed. Click on **OK**. The restaurants layer should also appear.

Runtime geodatabase

A Runtime geodatabase is basically a SQLLite database, which is ideal for a mobile app. It can contain feature layers for editing on a device. For more details on how to create this kind of geodatabase, see the ArcGIS Desktop help here: `http://desktop.arcgis.com/en/desktop/latest/map/working-with-arcmap/creating-arcgis-runtime-content.htm`.

Let's add a Runtime geodatabase to our app:

1. Add a `using` statement to `Esri.ArcGISRuntime.Data`.

2. Add the following method to the `ViewModel` class:

```
private async void CreateFeatureLayers()
{
    var gdb = await Geodatabase.OpenAsync(this.GDB);

    Envelope extent = null;
    foreach (var table in gdb.FeatureTables)
    {
        var flayer = new FeatureLayer()
        {
            ID = table.Name,
            DisplayName = table.Name,
            FeatureTable = table
        };

        if (!Geometry.IsNullOrEmpty(table.ServiceInfo.
            Extent))
        {
            if (Geometry.IsNullOrEmpty(extent))
                extent = table.ServiceInfo.Extent;
            else
                extent =
                    extent.Union(table.ServiceInfo.Extent);
        }
```

```
        this.mapView.Map.Layers.Add(flayer);
    }

    await this.mapView.SetViewAsync(extent.Expand(1.10));
}
```

3. Add this method at the end of the anonymous method of the `Messenger` class in the `ViewModel` constructor, as you have in previous steps with other methods. This method reads the Runtime geodatabase, creates `FeatureLayers`, adds them to the map, gets the extent of the `FeatureLayers`, and finally zooms around by 110 percent.

4. Run the app. You will see the parking meters of San Francisco. See the provided sample named `Chapter3a` if you need help.

Summary

In this chapter, you have been provided with a more thorough description of some of the concepts you were introduced to in earlier chapters. Namely, you have been provided with a description of the `MapView`, `Map`, and `Layers` classes along with some of the concepts required to understand projections and coordinate systems, scale, map interactions, and so on. You have also been introduced to the many kinds of layers that ArcGIS Runtime supports, which included an object model diagram that shows the breadth of supported layer types. Also, you were shown how to access online content while at the same time using `Behaviors`, which greatly expands the ability of your apps by providing a reference to the `MapView` class, so you can focus on requirements, instead of architecture. Lastly, you were introduced to adding offline content to a map.

Now that you've learned about the map and layer in 2D, we'll next turn our attention to 3D, where we will find that many of the concepts we've learned here apply to 3D. We'll learn about why we would want to use 3D, how to change from 2D to 3D, navigation in 3D, 3D layers, surface models, MVVM in 3D, and 3D content.

4
From 2D to 3D

In this chapter, we will explore the exciting new 3D capabilities of ArcGIS Runtime. 3D is a major new feature that will bring 3D viewing capabilities while at the desktop or in the field on a mobile device for the first time. At this point in time, 3D is only available in the .NET version of ArcGIS Runtime (Windows, Windows Store, and Windows Phone), but it is expected to be released to all flavors when the Quartz (next major release) version of ArcGIS Runtime is available. To start off, we'll see how easy it is to change a 2D map to 3D, and then we'll discuss how to navigate in 3D, adding layers to a 3D scene, creating surface models, and so on, all of which will be done with the MVVM pattern. We'll discuss the following topics:

- Why 3D?
- From 2D to 3D
- Navigating in 3D
- 3D layers
- Surface models
- MVVM in 3D
- 3D content

Why 3D?

Before we discuss how to change from 2D to 3D, it's important to understand why 3D is necessary. There are many reasons:

- Our world is of course 3D, so humans naturally are able to interact with a map in 3D in a more natural way than in 2D. 3D is more realistic than 2D, so it makes understanding a design or analytical output easier.

- For some applications of GIS, a 2D map is simply insufficient. If you're a military mission planner, 3D is just not a nice feature, it's a must. Imagine if you were trying to determine if an inbound flight had enough clearance to pass over some buildings. Without 3D, this would be impossible, so lives would be at risk.

- 3D allows urban planners to try out different designs before ever building anything. Using 3D allows the planner to see different possibilities to ensure other views aren't blocked or that a new building's shadow won't be cast on the other buildings using an artificial light source, which can be determined at different times of the year.

- 3D is also every helpful from an analytical perspective. Imagine modeling a flood in 2D. You'd be able to view the area of the flood but not the depth. Without the depth, we can't actually see how many floors of buildings would be inundated or whether a bridge would be covered. Another analytical example would be the use of 3D to determine whether the viewer can see something based on their **line of sight** (**LOS**).

- With 3D, improved modeling is possible. Maps have always been rough approximations of the Earth. With 3D, GIS, and new sensor technologies, such as **Light Detection and Ranging** (**LiDAR**), Radar, and **Synthetic Aperture Radar** (**SAR**), the ability to model the surface of the Earth has greatly improved, so it's now possible to make very detailed maps in 3D, which accurately show the position of objects in all three dimensions.

- With all of the sensors available in today's smart phones and tablets, 3D is the perfect environment to exploit them because these sensors can take advantage of the 3D nature of our world. For example, photographs or videos taken with smart phones are orientated and placed into their proper position in 3D. In addition, sensors allow us to measure the distance and height of objects so that we can correctly place them or even remove them from a scene.

- The **Augmented Reality** (**AR**) is about placing digital information on top of whatever you are viewing on your device. A perfect example of AR is a **head-up display** (**HUD**) used by pilots that allow them to access critical information, such as speed and heading, while seeing the real world. Another example is a utility field, where the user could point their phone around them and see underground water pipes in their exact position.

- The **Virtual Reality** (**VR**) toolkit allows a user to place themselves into the 3D scenes using head-mounted displays. Another military example of this is where VR helps soldiers train for battlefield scenarios. An example of this is the Synthetic Environment Core of the U.S. Army, where soldiers are able to navigate through virtual environments, as shown here:

Soldiers who navigate through viral environments

The preceding image is available at `http://www.peostri.army.mil/PRODUCTS/SECORE/images/2012_SECore-Dismount.gif`.

From 2D to 3D

Turning an ArcGIS Runtime app from 2D to 3D requires very little effort on your part. In fact, to go from 2D to 3D, you only have to know that there are equivalents of the `MapView` and `Map` classes in 3D. They are called the `SceneView` and `Scene` objects, respectively. Therefore, all that is is necessary is to change the code to this:

```
<esri:MapView x:Name="MyMapView>
    <esri:Map x:Name="theMap" >

    </esri:Map>
</esri:MapView>
```

The preceding code is to be changed to this:

```
<esri:SceneView x:Name="MySceneView"
    <esri:Scene x:Name="theScene" >

    </esri:Scene>
</esri:SceneView>
```

That's it. All that changed is that we used a `SceneView` control instead of `MapView` and a `Scene` control instead of `Map`. That was pretty easy, right? However, if you run the app with just the preceding XAML code, you'd only see this:

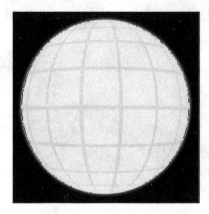

It's a globe, but it doesn't have any layers. The good news is that you can use all of the layers we've been using in 2D also. To do the same with code, we only need code similar to this:

```
// create a new SceneView
var mySceneView = new Esri.ArcGISRuntime.Controls.SceneView();

// create a new Scene
var myScene = new Esri.ArcGISRuntime.Controls.Scene();

var uri = new
    Uri("http://services.arcgisonline.com/arcgis/rest/services/
        World_Street_Map/MapServer");
var baseLayer = new ArcGISTiledMapServiceLayer(uri);
baseLayer.ID = "BaseMap";

// add the layer to the Scene
myScene.Layers.Add(baseLayer);

// add the Scene to the SceneView
mySceneView.Scene = myScene;

// add the SceneView to the grid
MyGrid.Children.Add(mySceneView);
```

Let's change the app we made in *Chapter 3*, *Maps and Layers*, and make it a 3D app:

1. Make a new ArcGIS Runtime project, install `Json.NET` and MVVM Light, and copy all code over from that project. Be sure to update all namespaces.

2. As you will recall, we made a custom behavior to change how the `MapView` class worked. Open `MapViewBehavior.cs` and change the following code in the class to this:

```
namespace Chapter4.Behaviors
{
    public class SceneViewBehavior : Behavior<SceneView>
    {
        protected override void OnAttached()
        {
            base.OnAttached();
            AssociatedObject.Loaded +=
                AssociatedObject_Loaded;
        }
        void AssociatedObject_Loaded(object sender,
            RoutedEventArgs e)
        {

            SceneView sceneView = sender as SceneView;
            Messenger.Default.Send<SceneView>(sceneView);
        }
    }
}
```

3. Rename the file from `MapViewBehavior.cs` to `SceneViewBehavior.cs`.

4. Update the XAML's `MapView` to `SceneView` properties and `Map` to `Scene`.

5. Update the XAML code in the `MainWindow.xaml` file of `SceneViewBehavior` like this:

```
<i:Interaction.Behaviors>
    <behave:SceneViewBehavior/>
</i:Interaction.Behaviors>
```

6. In the `MainViewModel.cs` file, update all references to the `MapView` private member to `SceneView` and change the reference of `MapView` to `SceneView`. Remove the `MapGrid` and `InteractionOptions` code. They don't work in 3D. Make sure that each reference to `this.mapView.Map` is now `this.sceneView.Scene`. Remove the method named `SetInitialExtent` because it won't work in 3D.

7. Comment out the `Zoom` method. Lastly, comment the `ZoomRelayCommand` and any other code related to it. We're going to bring these lines back in the next section.

8. Comment out the `Zoom` button line in the XAML code.

9. Run the app. You will see a virtual globe.

Congratulations! You've written your first 3D app. What's more, you created a 3D app that follows the MVVM pattern. While it's true that it's that simple, there are some other things we really need to understand before we move forward. For example, there are many significant differences between navigating in 3D, some differences in layers, how we set the Viewpoint, how we create geometry, and so on. So let's learn a little more before we build a more substantial 3D app.

Interacting with the Scene property

If you're new to 3D, the first challenge you will encounter is how to move around in the scene. Let's first cover these maneuvers with a mouse, so you can enhance your skills:

1. To pan with your mouse, hold the left mouse button and drag. You can also use the Arrow keys.

2. To zoom in, scroll the wheel forward on your mouse or hold the center button on your mouse if you have a three-button mouse. Use the + key to zoom in.

3. To zoom out, do the opposite of zooming in. Use the – key to zoom out.

4. To rotate, hold the right mouse button and drag.

You can't change the pitch, heading, or elevation with the mouse. However, you can do the following tasks with the keyboard:

1. To change the pitch, use the *W* (tilt up), *S* (tilt down), or *P* (perpendicular to the surface) keys.

2. To change the heading, use the *A* key to move west, the *D* key to move east, and the *N* key to set through north.

3. Lastly, with the keyboard, use the *U* key to move up and the *J* key to move down.

If you have a touch-screen device, you can of course use gestures:

1. To pan, use a single finger and drag, or you can flick.

2. To pan and zoom, use multiple fingers (spread zoom in, pinch zoom out, drag pans, and pivot rotates).

3. To change the heading and pitch, use multiple fingers (dragging up and down changes the pitch while dragging left or right changes the heading).

Lastly, to go into Fly mode, hold down *Q* and move the mouse in some direction, and this will allow you to pan and rotate based on the mouse position. As with driving a car, it takes a little getting used to, but before you know it, you'll master it.

Viewing in 3D

When you're in 3D, it is very much like being in the world. Your eye is like a camera looking at a target. 3D works in very much the same manner. However, as we are working with the entire Earth, we have to move the camera at high altitudes (elevation) to see anything. See the following screenshot:

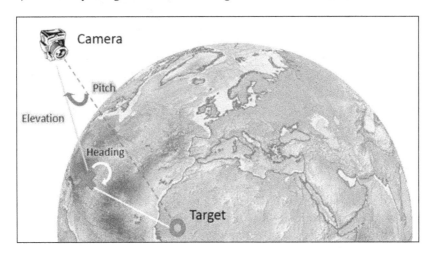

When viewing in 3D, you are the camera. The camera has an elevation above the surface of the Earth, a pitch, heading, and something it is looking at (target). This is always the case. The good news is that you can change any of these parameters by getting a reference to the camera of SceneView like this:

```
Camera camera = this.sceneView.Camera;
```

However, in order to do anything with the camera, we need to revisit the Viewpoint concept.

Viewpoints

As you saw with the `MapView` and `Map` classes, we could set the extent and scale, and then call `SetViewAsync` when we zoomed to Rome and everything worked perfectly. This is not the case with 3D. In 3D, there is no scale because we are dealing with the actual shape of the Earth in 3D. The scale is the same everywhere in 3D. Also, the extent doesn't apply either because the extent is a 2D concept. It's a rectangle laid on the flat Earth. With 3D, we're viewing from a camera. However, to make this easy to adjust and write code, Esri has transferred the Viewpoint concept over from 2D to 3D.

Let's add the `Zoom to Italy` button back to our code and XAML; but this time, replace the code with the following code:

```
public async void Zoom()
{
    // define a new camera over Italy
    var camera = new Camera(41.9, 12.5, 330.0, 180.0,
        73.0);

    // create a ViewPoint with the camera, a target
        geometry
    // (Envelope), and rotation (same as camera heading)
    var viewPoint = new Viewpoint(camera, new
        Envelope(12.5,
        41.85,12.6, 41.95, SpatialReferences.Wgs84),
            126.0);
    var animationDuration = new TimeSpan(0, 0, 2);

    await this.sceneView.SetViewAsync(viewPoint,
        animationDuration, true);
}
```

This new **Zoom to Italy** button now creates a new camera object with the latitude of 41.9, longitude of 12.5, altitude of 330, heading of 180, and of pitch of 73 degrees. The camera is looking due south. Next, a Viewpoint is created; but this time, we're using an envelope as a **viewing frustum**. See here:

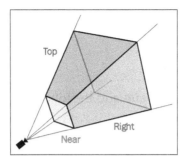

The envelope acts as a viewing frustum, which now sets a rectangle by which we're looking through out into infinite space, except the envelope has been set on its edge, as shown in the illustration. Lastly, we created a `TimeSpan` object and set it to 2 seconds, and then called `SetViewAsync` again with the Viewpoint animation duration, and set lift off to `true`, which makes the camera appear to lift off like a missile as it starts the zoom operation. Pretty nice, huh?

Surface model

As you zoom in and pan down to the surface of the sample app, you'll note that the Earth has no elevation relief. To remedy this, you will need a surface model of the surface of the Earth. A surface model has x and y coordinates just like other layers, but it also has z values for elevation. The z values are typically in meters or feet. At first, we'll use an online service, and then we'll use a file source for offline content.

Online services

As with the other services we've been using, Esri also has some services ready for us to use; but in order to do that, we need to update our model:

1. Open `Model.cs` and add the following `private` member:

    ```
    private string elevationLayerUri =
        "http://elevation3d.arcgis.com/arcgis/rest/services/
            WorldElevation3D/Terrain3D/ImageServer";
    ```

2. Also, add the following property:

    ```
    public string ElevationLayerUri
    {
        get { return this.elevationLayerUri; }
        set
        {
            if (value != this.elevationLayerUri)
    ```

```
        {
            this.elevationLayerUri = value;
        }
    }
}
```

3. Open `MainViewModel.cs` and add the following property:

    ```
    public string ElevationUri
    {
        get { return myModel.ElevationLayerUri; }
        set
        {
            this.myModel.ElevationLayerUri = value;
            RaisePropertyChanged("ElevationUri");
        }
    }
    ```

4. In `ConfigService`, change the line that reads the JSON file to read `config_scene.json`.

5. In the constructor of `MainViewModel`, where the layers are added to the `Messenger` code, add the following lines before the other layers:

    ```
    Uri uriElevation = new Uri(this.ElevationUri);
    ServiceElevationSource elevSource = new
        ServiceElevationSource(uriElevation);
    elevSource.IsEnabled = true;
    this.sceneView.Scene.Surface.Add(elevSource);
    ```

6. Update the `Zoom` method to the following code:

    ```
    public async void Zoom()
    {        {
        // define a new camera over Everest
        var camera = new Camera(28.1, 86.9253, 8000, 180,
            73.0);

        var viewPoint = new Viewpoint(camera, new
            Envelope(27.9, 86.7,
            28.1, 87.3, SpatialReferences.Wgs84), 126.0);
        var animationDuration = new TimeSpan(0, 0, 2);

        await this.sceneView.SetViewAsync(viewPoint,
            animationDuration,
            true);
    }
    ```

7. In `MainWindow.xaml`, change the button's name to `Zoom to Everest`.

8. Run the app and click on **Zoom to Everest**.

That's how easy it is to add a surface model to your scene. Note that in step 4, we set up the elevation source, enabled it, and then added it to the scene's surface. We now have terrain relief for the entire planet. Then, we updated the `Zoom` method to position the camera to look at `Everest`. For more information on creating and publishing this kind of service, navigate to `http://server.arcgis.com/en/ server/latest/publish-services/windows/publishing-image-services.htm`.

Offline sources (files)

It's also possible to set an elevation source using local files. ArcGIS Runtime supports the following elevation file formats:

- The **ARC Standard Raster Product (ASRP)** or **UTM/UPS Standard Raster Product (USRP)** format

- Controlled Image Base 1, 5, and 10 (`http://www.aviation.dla.mil/rmf/ products_digital.htm#4`)

- Digital Terrain Elevation Data 0, 1, and 2 (`https://en.wikipedia.org/ wiki/DTED`)

- GeoPackage (`http://www.geopackage.org/`)

- GeoTIFF (`http://trac.osgeo.org/geotiff/`)

- HFA

- HRE (`http://www.asprs.org/a/publications/proceedings/ sanantonio09/Heady.pdf`)

- The **Erdas Imagine (IMG)** format

- JPEG (`https://en.wikipedia.org/wiki/JPEG`)

- JPEG 2000 (`https://en.wikipedia.org/wiki/JPEG_2000`)

- NITF (`https://en.wikipedia.org/wiki/National_Imagery_ Transmission_Format`)

- PNG (`https://en.wikipedia.org/wiki/Portable_Network_Graphics`)

- RPF (`http://earth-info.nga.mil/publications/specs/ printed/89038/89038_CADRG.pdf`)

- The **Shuttle Radar Topography Mission (SRTM)** data 1 and 2 (`http:// www2.jpl.nasa.gov/srtm/`)

Let's use some elevation data that comes from the SRTM data:

1. Comment out the following lines:

```
Uri uriElevation = new Uri(this.ElevationUri);
ServiceElevationSource elevSource = new
    ServiceElevationSource(uriElevation);
elevSource.IsEnabled = true;
this.sceneView.Scene.Surface.Add(elevSource);
```

2. Replace them with these lines of code:

```
FilenameCollection fileCollection = new
    FilenameCollection();
fileCollection.Add(@"C:\ArcGISRuntimeBook\Data\SRTM\
    n27_e086_3arc_v2.dt1");
fileCollection.Add(@"C:\ArcGISRuntimeBook\Data\SRTM\
    n27_e087_3arc_v2.dt1");
fileCollection.Add(@"C:\ArcGISRuntimeBook\Data\SRTM\
    n28_e086_3arc_v2.dt1");
fileCollection.Add(@"C:\ArcGISRuntimeBook\Data\SRTM\
    n28_e087_3arc_v2.dt1");

FileElevationSource fileElevSource = new
    FileElevationSource();
fileElevSource.Filenames = fileCollection;
fileElevSource.IsEnabled = true;
fileElevSource.ID = "Elevation Source";
this.sceneView.Scene.Surface.Add(fileElevSource);
```

3. Run the app again and click on **Zoom to Everest**. The only difference you'll note is that the resolution is lower. The online service has a range of resolutions from 1,000 m to 3 m but the SRTM is 30 m resolution.

It should be noted that in a production-level app, you'd want to make a service factory that reads these files from disk, and then load them into your model.

Other properties of the SceneView control

Now that we've seen some ways of viewing elevation sources and layers in 3D, let's discuss some other properties of the SceneView control:

- IsShadowsEnabled: This property determines whether or not shadows appear in the scene.

- ElevationExaggeration: Using a factor, such as 1.5, this property determines the exaggeration of the surface model. The higher the number the more the scene is exaggerated.

- `LightPoint`: This property determines the position of the Sun using the x, y, and z values. It's recommended to use the `SetSunTime` method to make the light appear more realistic for the time of day.

- `AmbientLight`: This property sets the color of the ambient light in the scene when shadows are enabled.

3D content

It's one thing to see a basemap with a surface model, but what we really want to do is add content to the scene to add context. Let's make another change to our sample app:

1. Once again, change the `Zoom` method to the following code:

```
public async void Zoom()
    {
    // define a new camera over San Francisco
    var camera = new Camera(37.7833, -122.4167, 330.0, 45,
        73.0);
    // create a ViewPoint with the camera, a target
        geometry
    // (Envelope), and rotation (same as camera heading)
    var viewPoint = new Viewpoint(camera, new Envelope(37,
        -122, 38,
        -123, SpatialReferences.Wgs84), 126.0);
    var animationDuration = new TimeSpan(0, 0, 2);

    await this.sceneView.SetViewAsync(viewPoint,
        animationDuration,
        true);
}
```

2. Comment out the USA layer for now.

3. Change the name of the button in the XAML code to `Zoom to San Francisco`.

4. Add the following code near the end of the anonymous method of `Messenger`:

```
Uri sanFranciscoUri = new
    Uri("http://scene.arcgis.com/arcgis/rest/services/
        Hosted/Buildings_San_Francisco/SceneServer/layers/
            0");
SceneLayer sceneLayer = new SceneLayer(sanFranciscoUri);
sceneLayer.InitializeAsync();
this.sceneView.Scene.Layers.Add(sceneLayer);
```

5. Run the app and click on **Zoom to San Francisco**. You will be zoomed to this great city with the 3D textured buildings. Note that `SceneLayer` is specific to 3D and it won't work in `Map`.

You can find other cities at `http://scene.arcgis.com/arcgis/rest/services/Hosted`. Any of the cities that have `SceneServer` beside their name will have some 3D content. As in the preceding example, make sure to add `Layers/0` at the end of the service name.

Layering the Scene properties

We've now created a `SceneView` with a `Scene` property, added some basemaps, and the 3D cities, but we need to also talk about some of the properties of how these layers can be added. If you add `GraphicsLayer` to the `Scene` property, you also need to tell the layer how to place itself on the Earth. This is accomplished with `LayerSceneProperties`. You have three options:

- `Draped`: In this option, graphics will be drawn directly on the surface. The *z* values of the graphic are equal to the *z* value of the surface model.

- `Absolute`: With this option, the *z* values of the graphic are used independently of the surface model. If you want to show the altitude of an airplane, you would set the layer to use `Absolute`.

- `Relative`: Using this option means that the *z* value of the graphic would be added to the *z* value of the surface model. A good example of this would be the height of a power line above the terrain.

Here is an example of setting `LayerSceneProperties`:

```
var graphicsLayer = new GraphicsLayer();
graphicsLayer.ID = "DrapedGraphics";
graphicsLayer.SceneProperties.SurfacePlacement =
    SurfacePlacement.Draped;
MySceneView.Scene.Layers.Add(graphicsLayer);
```

With this code snippet, `GraphicsLayer` is instantiated, named, and `SurfacePlacement` is set to `Draped`. Any graphic added to this layer will appear on the surface.

Controlling the Scene example

Now that we've learned some important concepts regarding 3D, let's take our app and make it so the user can control the scene using an intuitive UI, while at the same time extending our understanding of MVVM by adding another `ViewModel` class to our app that makes the use of a reusable user control:

1. Create a new ArcGIS Runtime project, add MVVM Light, `Json.NET`, copy the code files from the previous project, and then update all references and the `using` statements.

2. Copy the XAML code from the `MainWindow.xaml` of `Chapter4` to the new `MainWindow.xaml` and update the `using` statements. Make sure the app runs as with the previous exercise. If you have any issues with the app not binding to the `ViewModel` class, check the binding in the XAML code and update them, like in this example:

   ```
   <TextBox Name="SearchTextBox" Text="{Binding SearchText,
       Source={StaticResource Locator}}"></TextBox>
   ```

3. Create a new MVVM Light `ViewModel` class. Be sure to select **MvvmViewModel (WPF)**. Call it `CameraViewModel.cs` and place it in the `ViewModels` folder. Add the following code to it:

   ```
   using GalaSoft.MvvmLight;
   using GalaSoft.MvvmLight.Messaging;

   using Esri.ArcGISRuntime.Geometry;
   using Esri.ArcGISRuntime.Controls;

   namespace Chapter4a.ViewModels
   ```

```
    {
        /// <summary>
        /// This class contains properties that a View can data
            bind to.
        /// <para>
        /// See http://www.galasoft.ch/mvvm
        /// </para>
        /// </summary>
        public class CameraViewModel : ViewModelBase
        {
            Esri.ArcGISRuntime.Controls.SceneView sceneView =
                null;
            double heading = 0.0;
            double pitch  = 0.0;
            double elevation = 11000000;
            double latitude = 0.0;
            double longitude = 0.0;
            /// <summary>
            /// Initializes a new instance of the
                LocationViewModel class.
            /// </summary>
            public CameraViewModel()
            {
                Messenger.Default.Register<SceneView>(this,
                    (sceneView) =>
                {
                    this.sceneView = sceneView;
                });
            }

            public double Heading
            {
                get { return this.heading; }
                set
                {
                    this.heading = value;
                    RaisePropertyChanged("Heading");

                    MapPoint myLocation = new
                        MapPoint(this.longitude,
                        this.latitude, this.elevation,
                            SpatialReferences.Wgs84);
                    Camera myCamera = new Camera(myLocation,
                        value, this.pitch);
                    this.sceneView.SetView(myCamera);
```

```
            }
        }
        public double Pitch
        {
            get { return this.pitch; }
            set
            {
                this.pitch = value;
                RaisePropertyChanged("Pitch");

                MapPoint myLocation = new
                    MapPoint(this.longitude,
                    this.latitude, this.elevation,
                        SpatialReferences.Wgs84);
                Camera myCamera = new Camera(myLocation,
                    this.heading, value);
                this.sceneView.SetView(myCamera);
            }
        }
        public double Elevation
        {
            get { return this.elevation; }
            set
            {
                this.elevation = value;
                RaisePropertyChanged("Elevation");

                MapPoint myLocation = new
                    MapPoint(this.longitude,
                    this.latitude, value,
                        SpatialReferences.Wgs84);
                Camera myCamera = new Camera(myLocation,
                    this.heading,
                        this.pitch);
                this.sceneView.SetView(myCamera);
            }
        }
        public double Latitude
        {
            get { return this.latitude; }
            set
            {
                this.latitude = value;
                RaisePropertyChanged("Latitude");
```

```
                    MapPoint myLocation = new
                        MapPoint(this.longitude, value,
                        this.elevation,
                            SpatialReferences.Wgs84);
                Camera myCamera = new Camera(myLocation,
                    this.heading,
                     this.pitch);
                this.sceneView.SetView(myCamera);
            }
        }
        public double Longitude
        {
            get { return this.longitude; }
            set
            {
                this.longitude = value;
                RaisePropertyChanged("Longitude");

                MapPoint myLocation = new MapPoint(value,
                    this.latitude,
                      this.elevation,
                          SpatialReferences.Wgs84);
                Camera myCamera = new Camera(myLocation,
                    this.heading,
                     this.pitch);
                this.sceneView.SetView(myCamera);
            }
        }
    }
}
```

4. In `ViewModelLocator`, register the new `ViewModel` class just like `MainViewModel` is registered:

```
SimpleIoc.Default.Register<CameraViewModel>();
```

5. Add a new property to `ViewModelLocator` like this:

```
public CameraViewModel MainViewModel
{
    get
    {
        return
            ServiceLocator.Current.GetInstance
                <CameraViewModel>();
    }
}
```

6. Create a new folder in your project and name it `UserControls`. Name the `UserControl` folder's `CameraUserControl.xaml` file like this:

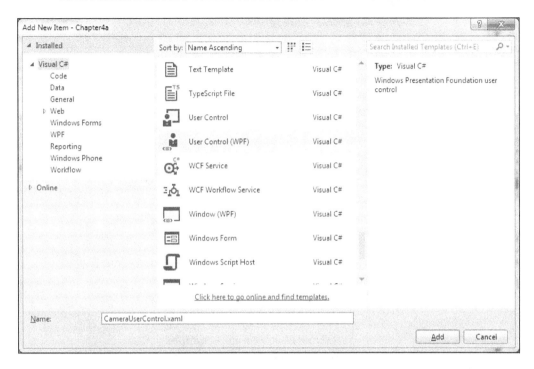

7. Add the following XAML code to `UserControl`:

```
<UserControl
    x:Class="Chapter4a.UserControls.CameraUserControl"
    xmlns="http://schemas.microsoft.com/winfx/2006/xaml/
        presentation"
    xmlns:x="http://schemas.microsoft.com/winfx/2006/xaml"
    xmlns:mc="http://schemas.openxmlformats.org/
        markup-compatibility/2006"
    xmlns:d="http://schemas.microsoft.com/expression/blend/
        2008"
    xmlns:i="clr-
        namespace:System.Windows.Interactivity;assembly=
            System.Windows.Interactivity"
    xmlns:vm="clr-namespace:Chapter4a.ViewModels"
        mc:Ignorable="d"
        d:DesignHeight="151" d:DesignWidth="240">

    <UserControl.Resources>
        <vm:CameraViewModel x:Key="Locator"
            d:IsDataSource="True"/>
```

```xml
        </UserControl.Resources>

<Grid DataContext="{Binding CameraViewModel,
    Source={StaticResource
    Locator}}">

    <StackPanel Orientation="Vertical" Margin="10">

        <!-- Slider for the Heading (in Degrees) of the
            Camera. -->
        <StackPanel Orientation="Horizontal">
            <Label Content="Heading: "
                Foreground="Green"
                FontWeight="Bold" Width="70" />
            <Slider x:Name="Slider_Heading" Width="100"
                AutoToolTipPlacement="BottomRight"
                    Minimum="0" Maximum="360"
                        TickFrequency="1"
                        Value="{Binding Heading,
                            Source={StaticResource
                            Locator}}"/>

            <TextBlock x:Name="Heading_Value"
                VerticalAlignment="Center"
                FontWeight="Bold"
                Foreground="Green"
                Text="{Binding Heading,
                    StringFormat='N0',
                    Source={StaticResource Locator}}"/>
        </StackPanel>

        <!-- Slider for the Pitch (in Degrees) of the
            Camera. -->
        <StackPanel Orientation="Horizontal">
            <Label Content="Pitch: " Foreground="Green"
                FontWeight="Bold"
                Width="70" />
            <Slider x:Name="Slider_Pitch" Width="100"
                AutoToolTipPlacement="BottomRight"
                Minimum="0" Maximum="180"
                    TickFrequency="1"
                    Value="{Binding Pitch,
                        Source={StaticResource
                        Locator}}"/>
```

```xml
        <TextBlock x:Name="Pitch_Value"
            VerticalAlignment="Center"
            FontWeight="Bold"
            Foreground="Green" Text="{Binding
                Pitch,
            StringFormat='N0',
            Source={StaticResource Locator}}"/>
    </StackPanel>

    <!-- Slider for the Z value (aka. Elevation in
        Meters) of the
        Camera's Location. -->
    <StackPanel Orientation="Horizontal">
        <Label Content="Elevation: "
            Foreground="Green"
                FontWeight="Bold" Width="70" />
        <Slider x:Name="Slider_Z" Width="200"
            AutoToolTipPlacement="BottomRight"
            Minimum="0" Maximum="15000000"
                TickFrequency="1"
                Value="{Binding Elevation,
                    StringFormat='N0',
                Source={StaticResource Locator}}"/>

        <TextBlock x:Name="Z_Value"
            VerticalAlignment="Center"
            Foreground="Green" Text="{Binding
                Elevation,
                    StringFormat='N0',
                    Source={StaticResource
                        Locator}}"/>
    </StackPanel>

    <!-- Slider for the X value (aka. Longitude in
        Decimal Degrees) of
            the Camera's Location. -->
    <StackPanel Orientation="Horizontal">
        <Label Content="Longitude: "
            Foreground="Green"
                FontWeight="Bold" Width="70"
        />
        <Slider x:Name="Slider_X" Width="200"
            AutoToolTipPlacement="BottomRight"
            AutoToolTipPrecision="2" Minimum="-180"
                Maximum="180"
                TickFrequency="0.01"
```

```
                        Value="{Binding Longitude,
                            Source={StaticResource
                        Locator}}" />

            <TextBlock x:Name="X_Value"
                Foreground="Green"
                FontWeight="Bold"
                    VerticalAlignment="Center"
                Text="{Binding
                    Longitude,StringFormat='N2',
                Source={StaticResource Locator}}"/>
        </StackPanel>

        <!-- Slider for the Y value (aka. Latitude in
            Decimal Degrees) of
            the Camera's Location. -->
        <StackPanel Orientation="Horizontal">
            <Label Content="Latitude: "
                Foreground="Green"
                FontWeight="Bold" Width="70" />
            <Slider x:Name="Slider_Y" Width="200"
                AutoToolTipPlacement="BottomRight"
                AutoToolTipPrecision="2" Minimum="-90"
                    Maximum="90"
                TickFrequency="0.01"
                Value="{Binding Latitude,
                    Source={StaticResource
            Locator}}"/>

            <TextBlock x:Name="Y_Value"
                Foreground="Green"
                FontWeight="Bold"
                    VerticalAlignment="Center"
                Text="{Binding Latitude,
                StringFormat='N2',Source=
                {StaticResource
                Locator}}"/>
        </StackPanel>


    </StackPanel>
  </Grid>
</UserControl>
```

8. In `MainWindow.xaml`, add the following `using` statement:

```
xmlns:uc="clr-namespace:Chapter4a.UserControls"
```

9. After the closing `SceneView` tag, add this line:

```
<uc:CameraUserControl HorizontalAlignment="Right"
    Margin="0, 50"> </uc:CameraUserControl>
```

10. Run the app. You will see a set of controls in the upper-right corner, as shown in the following screenshot:

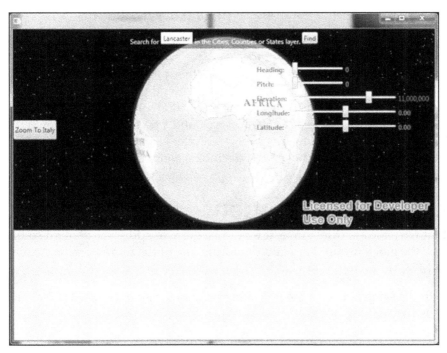

Display the set of controls in the upper-right corner

11. With these sliders, you can change the heading, pitch, elevation, longitude, and latitude. Note that the mouse isn't aware of these changes, so just use the sliders to change these parameters.

Several changes were made to this app to add these new capabilities. We created a new `ViewModel` class and added it to the MVVM Light registry of `ViewModel` classes. This required adding a property to ViewModel Locator so that the `View` class could find it. We then populated the new `ViewModel` class (`CameraViewModel`) with several properties so that the user control could bind to it. We then created a user control and set its `DataContext` to `ViewModel` so that it knows which one to use via the Locator. In this case, it's set to `CameraViewModel`.

In the user control, we bound each slider to a property on the ViewModel class. For example, the pitch slider was bound to the Pitch property so that when a user changes the pitch, it adjusts the globe's pitch. If you look at the pitch slider, you'll note that it has a range from 0 to 180, which is the range it should have. The latitude slider also has a range from -90 to 90 degrees, as it should be. The other sliders also have appropriate ranges. In the ViewModel class, the properties are updated every time the user changes a slider because RaisePropertyChanged is called for the respective property. Also, the camera is updated with the new position from a newly created MapPoint geometry. (We'll talk more about MapPoint and other geometries in the next chapter.) Lastly, the SceneView property is updated with this. sceneView.SetView(myCamera).

The really important new capability that we have created here is a reusable UserControl folder that we could put into another project if we needed it. All that would be necessary is we just copy the ViewModel and UserControl properties over into the project and update its Locator, and it should work perfectly fine. An even better solution would be to place UserControl and ViewModel into its own reusable class library, and then add it to any project. Of course, before this new user control could really be used, it would need to take into account the mouse movements.

Adding the 3D content

Now that we have a virtual globe, it would be great to add some content so that we can make the globe useful. Let's add data from one of the more popular formats, called KML:

1. In the same app, add the following code to the MainViewModel class after the USA layer:

    ```
    Uri kml = new
        Uri("http://earthquake.usgs.gov/earthquakes/feed/v1.0/
            summary/1.0_week_age_link.kml");
    KmlLayer kmlLayer = new KmlLayer(kml);
    kmlLayer.ID = "Earthquakes";
    kmlLayer.InitializeAsync();
    this.sceneView.Scene.Layers.Add(kmlLayer);
    ```

2. Run the app, and you should see recent earthquakes around the world along with a legend.

Now that we've added a KML layer, let's add a simple graphic to the map too:

1. Add using statements to `Esri.ArcGISRuntime.Symbology.`
   `SceneSymbology` and `Windows.Media.Colors` in the `MainViewModel.cs` file.
   Add the following code to a method in your `ViewModel` class, and then call
   the method from the anonymous messenger code:

```
private void DrawSphere()
{
    // create a new point (MapPoint); pass x, y, and z
       coordinates in
    // the constructor
    var point = new MapPoint(-122.4167, 37.7833,  6000);

    // create a graphics layer
    GraphicsLayer graphicsLayer = new GraphicsLayer();
    graphicsLayer.ID = "Graphics Layer";
    graphicsLayer.SceneProperties.SurfacePlacement =
        SurfacePlacement.Absolute;
    graphicsLayer.InitializeAsync();

    // creates a red sphere with a radius of 20,000 meters
    var sphereSym = new SphereMarkerSymbol();
    sphereSym.Color = Colors.Red;
    sphereSym.Radius = 20000;

    // create the graphic
    var graphic = new Graphic(point, sphereSym);
    graphicsLayer.Graphics.Add(graphic);

    // add the graphics layers to the scene
    this.sceneView.Scene.Layers.Add(graphicsLayer);
}
```

2. Next, let's add a 3D model and place it above the sphere we just added.
   Add this method:

```
private void DrawModel()
{
    ModelMarkerSymbol mms = new ModelMarkerSymbol();
    mms.SourceUri = "Data/M-14/M-14.obj";
    mms.Scale = 1000;

    SimpleRenderer sr = new SimpleRenderer();
    sr.Symbol = mms;
```

```
GraphicsOverlay graphicsOverlay = new GraphicsOverlay()
{
    RenderingMode = GraphicsRenderingMode.Dynamic,
    Renderer = sr,
    SceneProperties = new LayerSceneProperties()
    {
        SurfacePlacement = SurfacePlacement.Relative
    }
};

MapPoint mp = new MapPoint(-122.4167, 37.7833, 55000);

Graphic gm = new Graphic(mp);
graphicsOverlay.Graphics.Add(gm);
this.sceneView.GraphicsOverlays.Add(graphicsOverlay);
}
```

3. Run the app and zoom over San Francisco. You will see the sphere and a futuristic jet, which will look like this:

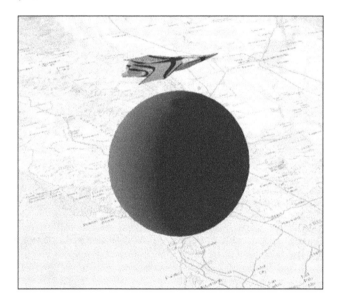

In this last example, there was a lot of new code you haven't seen yet. This is a taste of the topics to come in the next chapter, but this shows you how to add basic content to the scene. In the first method, a point was created via `MapPoint`, a new `GraphicsLayer` was created, and `SphereMarkerSymbol` was created with the color of red and a radius of 20,000 meters.

Lastly, the graphic was created using the point geometry (MapPoint) and a symbol, and then the graphic was added to GraphicsLayer. Then, GraphicsLayer was added to the Scene. In the second method (DrawModel), ModelMarkerSymbol was created, its SourceUri property was set, SimpleRenderer was instantiated, and its symbol was set to ModelMarkerSymbol. Then, GraphicsOverlay was created with several properties, the geometry was set to be over San Francisco, the graphic was created with the geometry, the graphic was added to GraphicsOverlay, and finally, GraphicsOverlay was added to the GraphicsOverlay property of SceneView.

Although we created this with a simple method in ViewModel, it really should go into a class that allows you to reuse it in all of your projects. You could then have a service that your ViewModel class uses to call the DLL.

Summary

In this chapter, you've been introduced to 3D by taking a 2D app and converting it to 3D with very little effort. Not only that, you were also shown how to do this with the MVVM pattern. You were also shown how to interact with the scene, the differences between 2D and 3D viewpoints, how to use online and offline surface models, controlling the scene, and lastly, how to add some basic 3D content. You were also shown how to incorporate another ViewModel class into the overall design so that you can reuse it by having ViewModel bound to a WPF user control. With this information, you now have enough background knowledge to continue on to the next level so that you can create geometry and symbolize your own content.

In the next chapter, we will take a deeper dive into the ArcGIS Runtime API by looking at geometry and symbology. Learning about geometry and symbology will allow us to construct the 2D and 3D objects and make them appear in a wide variety of ways to meet user requirements.

5
Geometry and Symbology

In the previous chapters, we've been looking at ArcGIS Runtime at a high level. It's time to go at this from a different angle. In this chapter, we're going to go down to the lowest level of the ArcGIS Runtime and discuss geometry so that we can build any kind of shape, and then we're going to discuss how we display the geometry by setting up the symbology. We will discuss how to make geometrical objects from scratch and how to apply them to entire layers, whether they come from online services or data stored locally on disk. We'll discuss the following topics:

- Assigning geometry
- Geometric objects
- Geometry builders
- Units
- Geometry engine
- Simple symbols
- Scene symbols
- Military symbols
- Renderers
- Limitations

Assigning geometry

Before we discuss geometry, we need to talk about what you assign geometry to. Basically, you are going to assign geometry to a graphic or a geodatabase feature.

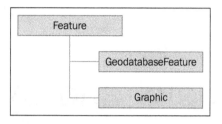

As you can see in the diagram, the GeodatabaseFeature and Graphic classes inherit from the abstract class called Feature. A Feature class has geometry and attributes (fields) and comes from either a feature service (online), Runtime geodatabase (offline), or some other source, such as a shapefile. A graphic can be added to GraphicsLayer or GraphicsOverlay. The big difference between a Graphic and a Feature class is that a Feature class is persisted to FeatureLayer, either in a web service or to disk. A Graphic class is stored in GraphicsLayer or GraphicsOverlay, but is not persisted. Both the GeodatabaseFeature and Graphic classes can have their geometry set like this:

```
// Create a new Graphic.
Graphic graphic = new Graphic();

// Create a new MapPoint and set it to the Graphics geometry.
MapPoint mapPoint = new MapPoint(0, 0, 10000,
    Geometry.SpatialReferences.Wgs84);
graphic.Geometry = mapPoint;
```

As shown here, a Graphic class is created along with a MapPoint geometry. The MapPoint geometry is then assigned to the Geometry property of Graphic, which allows the graphic to be placed in its correct location. Once that is done, it can be added to GraphicsLayer or GraphicsOverlay. Geometry can also be assigned to a GeodatabaseFeature class in a very similar fashion, but this is typically done interactively using the editor of MapView.

Geometric objects

In order to add any kind of geometry, such as a line, circle, or 3D fence, we must get down to the basics and discuss geometry in enough detail so that you are able to add more content to your map or scene, either by constructing the geometry in code or allowing the user to interactively create the geometry on the map or scene. (We will discuss interactive editing in *Chapter 9, Editing Features*.) In this section, we're going to consider the following object diagram and discuss each type of geometry:

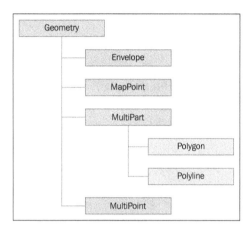

All objects in this section will come from `Esri.ArcGISRuntime.Geometry`.

Immutability

All of these geometries are immutable, which means that once you create them, you can't change them. However, all of these geometries can be created and modified with their respective builder classes. We will discuss the builder classes after we've introduced these geometries.

Geometry object

The `Geometry` object is an abstract class that defines a geometric shape, which means all subclasses inherit its properties and methods. The `Geometry` object has some important methods, such as `FromJson`, which allows you to convert from a JSON representation of the geometry. There is also `ToJson`, which allows you to convert a `Geometry` object to JSON. The `IsEqual` method will tell you whether two geometries are equal. By equal, we mean that their spatial references are the same. This also means that for geometries containing points, the order in which the points are created must also match. The `HasZ` property is a Boolean that simply lets you determine whether the object has the *z* values.

Lastly, the `Dimension` property tells you the dimensionality of the object. A point has zero dimensions. What this means is that if you were to put a bounding box around the object, the box itself would have an area of zero. A line has a dimensionality of 1 because if it starts at `(0,0)` and goes to `(0,1)`, and a box will have a width but not a height. Polygons are two dimensional. Volumes are three dimensional.

MapPoint

A `MapPoint` class is a fundamental geometry that you need to master. As you've seen in previous chapters, it's easy to create:

```
MapPoint point = new MapPoint(0,0);
```

A `MapPoint` class has six overloads on its constructor. To create a 3D point, just provide the *z* value:

```
MapPoint p1 = new MapPoint(0, 0, 0);
```

However, these simple examples don't say anything about the coordinate system we're using, so there is another overload that allows you to specify it, like this:

```
MapPoint point = new MapPoint(0, 0, 0, SpatialReferences.Wgs84);
```

By specifying WGS84, this point will be at latitude 0 degrees and longitude 0 degrees. This location is a very different location if the spatial reference is Mercator. To convert from one coordinate system to another, you will use the `GeometryEngine` method, which we will discuss later.

If you have lots of points that need to be created, a better method is to use a `PointCollection` class. For example, if you are loading in a CSV file with thousands of points, you could have code like this:

```
// create an empty point collection
var points = new PointCollection(SpatialReferences.WebMercator);

// set the capacity if you know the maximum number of points to
    add
points.Capacity = maxNumberOfPoints;

// repeat adding vertex coordinates directly into the
// PointCollection (no MapPoint instances)
while(lotsOfPoints)
{
    // ... read x/y coordinate values ...

    points.AddPoint(x,y);
```

```
}

// pass the point collection to the Polygon constructor
var polygon = new Polygon(points);
```

Envelope

An envelope is just a rectangle that you can use in multiple ways. As we discussed earlier, you can use it to set or get the map extent, the viewing frustum in 3D, or get the extent of a set of geometries. And envelope has three overloaded constructors. You can create an envelope with two `MapPoint` classes by setting the corners, using four coordinates, and by setting the four coordinates of the box along with a spatial reference.

Envelopes have several useful methods, such as `GetCenter`, `Intersect`, `Intersects`, and `Union`. These methods are useful when navigating the map or scene. For example, if your app has two maps and you want to merge the maps' two envelopes, you can do that with `Union`.

Multipart

Multipart is an abstract class for `Polygon` and `Polyline`. The most useful property in this class is parts. For example, if you have a polyline, you can get all of the constituent parts that make it up.

Polyline

A polyline is a collection of one or more segments (lines). As with `MapPoint`, this kind of geometry can be created with code or interactively via the mouse, touch, or stylus. To create a `Polyline` constructor, you have several options. The polylines can be created with `MapPoint`, an enumeration of `MapPoint` classes, an enumeration of `Segment` instances, or an enumeration of an enumeration of segments. For example, here's some code that creates a line from two `MapPoint` classes that are added to a generic `List` field:

```
MapPoint point1 = new MapPoint(0, 0, 0);
MapPoint point2 = new MapPoint(0, 1, 0);

List<MapPoint> list = new List<MapPoint>();
list.Add(point1);
list.Add(point2);

Polyline polyline = new Polyline(list, SpatialReferences.Wgs84);
```

To create `Polyline` using `LineSegment`, you can use the following code:

```
LineSegment linSegmentLeft = new LineSegment(point1, point2);
LineSegment linSegmentTop = new LineSegment(point2, point3);
LineSegment linSegmentRight = new LineSegment(point3, point4);
LineSegment linSegmentBottom = new LineSegment(point4, point1);

List<LineSegment> listOfLines = new List<ineSegment>();
Polyline pl = new Polyline(listOfLines);
```

Polygon

A polygon is a closed shape defined by three or more parts. Each part is a connected sequence of `Segment` instances. To create a polygon, the first point of the first `Segment` instance must be at the same location at the last point of the last `Segment` instance. This forms a closed ring. A polygon can have multiple rings, but they should not overlap. Here is a simple example of a polygon with one ring:

```
MapPoint point1 = new MapPoint(0, 0, 1);
MapPoint point2 = new MapPoint(0, 1, 1);
MapPoint point3 = new MapPoint(1, 1, 1);
MapPoint point4 = new MapPoint(1, 0, 1);

List<MapPoint> list = new List<MapPoint>();
Polygon polygon = new Polygon(list);
```

Here is another example of a polygon that contains two rings:

```
// interior ring
List<MapPoint> exteriorRing = new List<MapPoint>();
exteriorRing.Add(new MapPoint(5, 5));
exteriorRing.Add(new MapPoint(5, 10));
exteriorRing.Add(new MapPoint(10, 10));
exteriorRing.Add(new MapPoint(10, 5));

//exterior ring
List<MapPoint> interiorRing = new List<MapPoint>();
interiorRing.Add(new MapPoint(6, 6));
interiorRing.Add(new MapPoint(9, 6));
interiorRing.Add(new MapPoint(9, 9));
interiorRing.Add(new MapPoint(6, 9));
```

```
// List of rings
List<List<MapPoint>> myRings = new List<List<MapPoint>>();
myRings.Add(interiorRing);
myRings.Add(interiorRing);

Polygon myPolygon = new Polygon(myRings);
```

Once the polygon has been symbolized, it will look like this:

It's very important that the points that make up the polygon or polyline must be ordered. One other method worth noting about a polygon is that you can get a polyline representation using `Polygon.ToPolyline`. Also, a useful property is `IsEmpty`, which can be used to check whether you can perform an operation on the polygon. If the polygon contains no parts, you won't be able to do anything with it.

Multipoint

A `Multipoint` object is a single representation of multiple non-connected points. An example of this kind of geometry might be a collection of bore holes for a mining operation. Instead of tracking every point, they are grouped together to form a single record. Another example is a LiDAR point cloud. A `Multipoint` object has two overloaded constructors. The first one allows for an IEnumerable collection, and the other one also allows for an `IEnumerable` method with a `SpatialReference` instance. A `Multipoint` object is created like this:

```
MapPoint point1 = new MapPoint(0, 0, 1);
MapPoint point2 = new MapPoint(0, 1, 1);
MapPoint point3 = new MapPoint(1, 1, 1);
MapPoint point4 = new MapPoint(1, 0, 1);

List<MapPoint> list = new List<MapPoint>();
list.Add(point1);
list.Add(point2);
list.Add(point3);
```

```
    list.Add(point4);

Multipoint mp = new Multipoint(list);
// Once it is created you can access the individual MapPoints like
// this:
Multipoint multiPoint = new Multipoint(list);
for (int i = 0; i < multiPoint.Points.Count; i++)
{
    MapPoint point = multiPoint.Points[i];
}
```

Geometry builders

Now that you've seen how to create immutable geometries, let's go over the other ways of creating them so that they are mutable (they can be changed after they've been created). The primary reason you'd use these builders is so that a user can change the geometry interactively by adding, inserting, or removing parts of a polygon, for example, or by using a mouse, finder, or pen. Other than that, the class hierarchy is the same as shown in class diagram, in the *Geometric objects* section. However, as you can change these geometries, there are some other properties and methods. For example, `MapPointBuilder` has a method named `SetValue` that takes in x and y, and optionally, z values, so that you can change a point (vertex) of a polygon. The `MultipartBuilder` class allows you to get and set `PartCollection` that makes up the `Multipart` class. A `Multipart` class allows you to only get `RealOnlyPartCollection`. Also, `MultipartBuilder` has several methods, such as `AddPart` and `AddPoint`, which allow you to add geometries to it.

Units

Mastering units is vital for working with geographic data. Your users will enter data in one unit without realizing that the map is based on another unit. For example, many users in the U.S. expect to enter units in the U.S. feet, but the map may be in meters. The following diagram shows the object model diagram of the different kinds of unit you can convert. An abstract class named `Unit` provides the basis for instantiable classes named `LinearUnit`, `AreaUnit`, and `AngularUnit`:

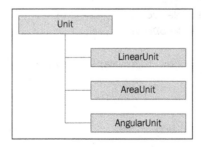

The `Unit` class has one method of interest, called `Create`, which allows you to create a unit using an ID that is a property of `Unit`. The `Unit` class also has the `Name` and `UnitType` methods. The good news is that you don't have to know these IDs or names. You can just use an ID like this:

```
AreaUnit squareYards =
    (AreaUnit)AreaUnit.Create(AreaUnits.SquareYards.Id);
```

In this case, the ID is 109442 and the name is `SquareYards`. For example, to convert from square yards to square meters, you would simply call this line of code after the preceding one:

```
double squareMeters = squareYards.ConvertToSquareMeters(100.0);
```

100 square yards would then be converted into square meters.

The `LinearUnit` class works in a similar fashion; in this class, you can convert distances to and from other units. The `AngularUnit` class allows you to convert angular units, such as degrees to radians and vice versa.

The `LinearUnit` class allows you to convert feet, inches, kilometers, meters, miles, millimeters, nautical miles, and yards. The `AreaUnit` class allows you to convert acres, areas, hectares, square centimeters, square decimeters, square feet, square inches, square kilometers, square miles, square millimeters, and square yards. The `AngularUnit` class allows you to convert degrees, grads, minutes, radians, and seconds.

Converting coordinates

When you work with layers in one coordinate system and need to convert them to another coordinate system, the `ConvertCoordinate` static class is your friend. With this class, you can convert back and forth between the **Global Area Reference System (GARS)** of **National Geospatial-Intelligence Agency (NGA)**, latitude/longitude, **Military Grid Reference System (MGRS)**, **U. S. National Grid (USNG)**, and **Universe Transverse Mercator (UTM)**. GARS and MGRS are particularly relevant to the military. For more information on GARS, UTM, MGRS, and USNG, navigate to `http://earth-info.nga.mil/GandG/coordsys/grids/referencesys.html`.

Here is an example of converting a latitude/longitude to MGRS:

```
MapPoint mapPointLatLon = new MapPoint(0, 0,
    SpatialReferences.Wgs84);
string coord = ConvertCoordinate.ToMgrs(mapPointLatLon,
    MgrsConversionMode.Automatic, 2, false, false);
```

The resulting string will contain a value that looks like this: `31NAA6600`. Another example illustrates converting from decimal degrees to degrees, minutes, and seconds:

```
MapPoint mapPointLatLon = new MapPoint(-177.23, 45.884,
    SpatialReferences.Wgs84);
string coord =
    ConvertCoordinate.ToDegreesMinutesSeconds(mapPointLatLon, 5);
```

Geometry engine

Once you have a geometry either from one constructed with code, as shown in the previous section, or one from clicking on (selecting) it, you can now perform geometric operations to derive other geometries by using the `GeometryEngine` class. What's powerful about this class is that it allows you to perform these operations in a connected and disconnected setting. It will also calculate the area or length of geometries such as polygons or polylines, respectively. With that said, there are too many options to describe, so we'll illustrate just a couple of them to give you a taste of what's possible.

Project

The `Project` method is one of the most important geometrical operations you will need from `GeometryEngine`, because it allows you to convert coordinates from one coordinate system to another. Users often want to enter coordinates in one coordinate system, but the basemap or layer in which the data will be stored will be in another coordinate system. As a result, you will need to re-project these coordinates. Also, in many apps, the user will enter data in one coordinate system, but the dialog they entered the data into will need to convert the units into another. Here is a typical use of this method:

```
MapPoint mapPointLatLon = new MapPoint(-80.845504, 35.211432,
    SpatialReferences.Wgs84);

var webMercatorSpatialRef = new Geometry.SpatialReference(102100);
var webMercatorPoint = GeometryEngine.Project(mapPointLatLon,
    webMercatorSpatialRef);
```

In this example, a set of coordinates in latitude/longitude have been converted to `WebMercator`. This is a very common projection for Web maps and the basemaps of Esri.

Buffering

Buffering is an analytical operation that allows you to do proximity analysis. For example, you can take a `MapPoint` class and buffer it to derive a polygon, which can then be used to select everything in a layer that touches or overlaps this buffer. Here's a code sample:

```
MapPoint mapPointLatLon = new MapPoint-80.845504, 35.211432,
    SpatialReferences.Wgs84);

var webMercatorSpatialRef = new
    Esri.ArcGISRuntime.Geometry.SpatialReference(102100);
var myWebMercPoint = GeometryEngine.Project(mapPointLatLon,
    webMercatorSpatialRef);

Geometry buffer = GeometryEngine.Buffer(myWebMercPoint, 10000);
```

In this code, we've created a `MapPoint` class using latitude/longitude, re-projected it to `Mercator`, and then buffered the point with a distance of 10,000 meters. We needed to re-project the `MapPoint` class because we needed to use a linear distance in meters. Once that is done, we then buffered it and produced a geometry. A buffer always creates a resulting polygon, so it can be cast to polygon. Here's a nice diagram from Esri (`https://developers.arcgis.com/net/desktop/guide/GUID-0D7163BF-1724-414F-A58C-F3361B648B36-web.png`) that shows what happens when you buffer polygons, polylines, and points.

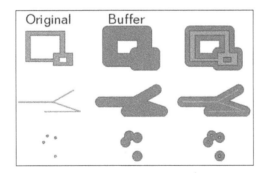

Once you have a buffer, you can perform other operations on it using `GeometryEngine`, such as measure its area like this:

```
double polyArea = GeometryEngine.Area(buffer);
```

The `GeometryEngine` method contains the following operations that you will need to investigate to solve the problem you are trying to solve. Here's a complete list of operations:

- `Area`
- `AutoComplete`
- `Boundary`
- `Buffer`
- `Clip`
- `Contains`
- `ConvexHull`
- `Crosses`
- `Cut`
- `Densify`
- `Difference`
- `Disjoint`
- `Distance`
- `Equals`
- `Extend`
- `Generalize`
- `GeodesicArea`
- `GeodesicBuffer`
- `GeodesicDensify`
- `GeodesicDistance`
- `GeodesicEllipse`
- `GeodesicLength`
- `GeodesicMove`
- `GeodesicSector`
- `Intersection`
- `Intersects`
- `LabelPoint`
- `Length`
- `NearestCoordinate`
- `NearestVertex`
- `Offset`

- Overlaps
- Project
- Relate
- Reshape
- ShapePreservingArea
- ShapePreservingLength
- SymmetricDifference
- Touches
- Union
- Within

It should be noted that the preceding buffer used Euclidean distance, which would not be accurate for large areas on a globe because the Earth is an ellipsoidal shape. To do this correctly, use `GeodesicBuffer` because it will account for the actual shape of Earth. An Euclidian buffer works fine for small areas, such as when all of the geometries are within one UTM zone. For larger areas, use `GeodesicBuffer`. Here's an example of the differences, which can also be seen at `http://www.esri.com/news/arcuser/0111/graphics/geodesic_6.jpg`:

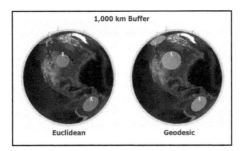

Difference in the buffers

As shown in the preceding screenshot, the **Geodesic** buffer is larger, and in this case accurate because it covers a larger area of the surface of Earth. These kinds of difference also apply to polyline geometry.

Symbology

Now that you've had a taste of how to create, change, and operate on geometry, it's time to discuss how we make the geometry appear on the map or scene. Symbology is part of cartography, which is both an art and science. ArcGIS Runtime has the ability to symbolize all of the geometries we have just discussed using cartographic principles, because it is essentially a GIS. With ArcGIS Runtime, you can make the map appear in such a manner that is visually pleasing, while at the same time scientifically correct.

After the geometry is created and validated, you will need to consider the following visual variables for quantitative data: Spacing, Size, Perspective, Height, and Color (hue, lightness, and saturation). For qualitative data, it's necessary to consider Orientation, Shape, Arrangement, and Color (hue). ArcGIS Runtime has the ability to consider all of these when you assign a symbol to a feature or graphic. First, we will go over simple symbols for the geometry we created earlier. Then, we'll discuss how to render entire layers, such as the FeatureLayer class. Finally, we'll discuss special kinds of symbol for military symbology.

Simple symbols

For simple symbols, we will utilize Esri.ArcGISRuntime.Symbology. In the examples that follow, we will take some of the geometries we used earlier, create a graphic, and then assign the symbol and geometry to the graphic. Once the graphic is created, we will then add it to GraphicsLayer or GraphicOverlays so that it will be shown. We can also assign geometry to a Feature class, and then add that to FeatureLayer, but we will reserve editing the FeatureLayer class for *Chapter 9, Editing Feature* (Online and Offline). Here is the symbology object model diagram:

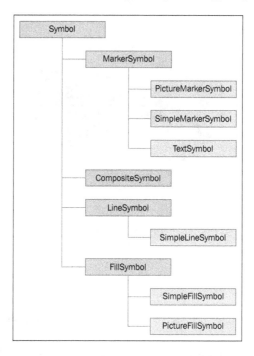

Symbol

Symbol is an abstract class that all subclasses inherit from. From this point, each kind of symbol under Symbol applies to certain kinds of geometry. For example, a MarkerSymbol subclass applies to MapPoint, LineSymbol applies to the Polyline geometry, and FillSymbol applies to the Polygon geometry. A Symbol class has some useful methods, such as CreateSwatchAsync, which returns an image of the symbol. It has three overloads. Just like Geometry, there's also a FromJson and a ToJson method.

MarkerSymbol

A MarkSymbol subclass is a subclass of Symbol and a base class of PictureMarkerSymbol, SimpleMarkerSymbol, and TextSymbol. The MarkerSymbol subclass has four properties worth noting: Angle, AngleAlignment, XOffset, and YOffset. With these properties, you can set the angle of the symbol, the angle alignment type, and the offset of the symbols in *x* and *y* coordinates. With AngleAlignment, you have the option to set whether the MarkerSymbol subclass rotates depending on how GraphicsLayer is configured with GraphicsLayer. RenderingMode.

SimpleMarkerSymbol

The most basic kind of marker symbol you will apply to point geometry is the SimpleMarkerSymbol class. With this class, you can set the color, outline color, size, and style. Here's an example of creating a Graphic class, creating a MapPoint class, setting MapPoint to Graphic, creating SimpleMarkerSymbol, setting that to the Graphic class, and then adding the Graphic class to the GraphicsLayer class, which is then added to Scene:

```
// create a graphics layer
GraphicsLayer graphicsLayer = new GraphicsLayer();
graphicsLayer.ID = "My Layer";
graphicsLayer.InitializeAsync();

// Create a new Graphic.
Graphic myGraphic = new Graphic();

// Create a new MapPoint using the data from the text file and set
    it // to the Graphics geometry.

MapPoint myMapPoint = new MapPoint(0, 0, 10000,
    Geometry.SpatialReferences.Wgs84);
myGraphic.Geometry = myMapPoint;
```

```
// Create the symbol
SimpleMarkerSymbol myMarkerSymbol = new SimpleMarkerSymbol();
myMarkerSymbol.Size = 25;
myMarkerSymbol.Color = Colors.Red;
myMarkerSymbol.Style = SimpleMarkerStyle.Diamond;
myGraphic.Symbol = myMarkerSymbol;

// Add the Graphic to the GraphicLayer.
graphicsLayer.Graphics.Add(myGraphic);
this.sceneView.Scene.Layers.Add(graphicsLayer);
```

The `Style` class can be set using `SimpleMarkerStyle`, which has these options: circle, cross, diamond, square, triangle, and an X.

PictureMarkerSymbol

A `PictureMarkerSymbol` class is used to add a picture on a graphic's or feature's point geometry. The most important methods in this class are `SetSource` and `SetSourceAsync`. An example of its use is provided here:

```
System.Uri myPictureUri = new
    System.Uri("http://static.arcgis.com/images/Symbols/
        Transportation/SkullandCrossbones.png");

// Create a new PictureMarkerSymbol based on the static image and
    set // a few properties.
PictureMarkerSymbol pictureMarkerSymbol = new
    PictureMarkerSymbol();
pictureMarkerSymbol.SetSourceAsync(myPictureUri);
pictureMarkerSymbol.Angle = 30;
pictureMarkerSymbol.Height = 30;
pictureMarkerSymbol.Width = 30;
```

In this example, the image comes from an online image source, but it could easily have come from an image in your Visual Studio project.

TextSymbol

The `TextSymbol` class allows you to place text on a graphic or feature. Of interest with this class are `Font` and `Text`. Obviously, you need to give the `TextSymbol` class a string with the `Text` property, but you also need to provide a font. Here is an example using a graphic:

```
TextSymbol textSymbol = new TextSymbol();
textSymbol.Text = "Some Text";
textSymbol.Color = System.Windows.Media.Colors.Black;
```

```
SymbolFont symbolFont = new SymbolFont();
symbolFont.FontFamily = "Courier New";
symbolFont.FontSize = 25;
symbolFont.FontStyle = SymbolFontStyle.Normal;
symbolFont.FontWeight = SymbolFontWeight.Bold;
textSymbol.Font = symbolFont;

Graphic graphic = new Graphic();
graphic.Geometry = myMapPoint;
graphic.Symbol = textSymbol;
graphicsLayer.Graphics.Add(graphic);
```

LineSymbol

The `LineSymbol` class is a base class that allows you to apply symbols to polylines and the outlines of polygon features and graphics. It contains one subclass called `SimpleLineSymbol`, which can be used like this:

```
SimpleLineSymbol sls = new SimpleLineSymbol()
{
    Color = Colors.Black, Style = SimpleLineStyle.Solid, Width = 2
};
```

FillSymbol

The `FillSymbol` class is a base class of `SimpleFillSymbol`. It is implemented like so:

```
SimpleFillSymbol sfs = new SimpleFillSymbol()
{
    Color = Colors.Red,
    Style = SimpleFillStyle.Solid
};
```

With this symbol, you can set the style to `BackwardDiagonal`, `Cross`, `DiagonalCross`, `ForwardDiagonal`, `Horizontal`, `Null`, `Solid`, and `Vertical`. You can also set the outline of the polygons using a `SimpleLineSymbol`, by way of the `Outline` property.

CompositeSymbol

As the name implies, `CompositeSymbol` allows you to combine symbols of any kind to form a new symbol. Here is an example:

```
CompositeSymbol compositeSymbol = new CompositeSymbol();
compositeSymbol.Symbols.Add(markerSymbol);
compositeSymbol.Symbols.Add(textSymbol);
```

Scene symbols

For 3D, there are several additional symbols in `Esri.ArcGIS.Runtime.SceneSymbology`. At the end of the previous chapter, we discussed `ModelMarkerSymbol` but didn't provide any details on how it works. As a refresher, here's the code snippet:

```
ModelMarkerSymbol mms = new ModelMarkerSymbol();
mms.SourceUri = "Data/M-14/M-14.obj";
mms.Scale = 1000;
```

The first thing that you have to do is set the source, which in this case is a `Wavefront Object` file, using a relative path. The `Data` directory must be located where your `.exe` file is located. The `ModelMarkerSymbol` class supports the following 3D model formats:

- Collada (`.dae`)
- Blender 3D (`.blend`)
- 3ds Max 3DS (`.3ds`)
- 3ds Max ASE (`.ase`)
- Wavefront Object (`.obj`)
- Industry Foundation Classes (IFC/step) (`.ifc`)
- XGL (`.xgl`, `.zgl`)
- Stanford Polygon Library (`.ply`)
- AutoCAD DXF (`.dxf`)
- LightWave (`.lwo`)
- LightWave Scene (`.lws`)
- Modo (`.lxo`)
- Stereolithography (`.stl`)
- DirectX X (`.x`)
- AC3D (`.ac`)
- Milkshape 3D (`.ms3d`)
- TrueSpace (`.cob`, `.scn`)

You can also create the simple 3D marker symbols with `BoxMarkerSymbol`, `ConeMarkerSymbol`, `DiamondMarkerSymbol`, `SphereMarkerSymbol`, and `TetrahedronMarkerSymbol`. For lines, you can create `SceneLineSymbol` and `TubeLineSymbol`. All of these symbols inherit from `SceneSymbol`. The 3D marker symbols allow you to set the heading, pitch, and roll. Here is an example of `BoxMarkerSymbol`:

```
BoxMarkerSymbol box = new BoxMarkerSymbol();
box.Width = 10000;
box.Height = 10000;
box.Depth = 10000;
```

Renderers

So far, we've been creating symbols for single graphics, but if you have a `FeatureLayer` class with thousands or even millions of records (features) in it, you obviously won't create a graphic for each object you want on the map. A renderer allows you to apply the same symbol to all features in the layer in one go, and it is much faster. See the following diagram:

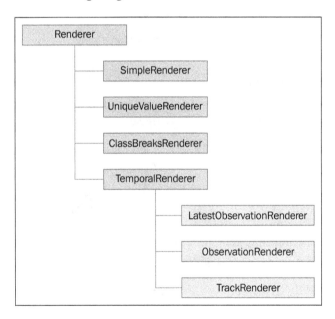

The preceding diagram shows the object model diagram for renderers. The `Renderer` class is the base class and like other abstract classes in this chapter, it has the ability to convert to and from JSON. It also has the ability to set `SceneProperties`.

SimpleRenderer

If you want to apply the same symbol to every feature in the layer, `SimpleRenderer` is the solution. Here is an example that will show the locations of all the offices of Esri around the world:

```
Uri uri = new
    Uri("https://services.arcgis.com/P3ePLMYs2RVChkJx/ArcGIS/rest/
        services/Esri_Offices/FeatureServer/0");

FeatureLayer featureLayer = new FeatureLayer(uri);
featureLayer.InitializeAsync();

SimpleMarkerSymbol markerSymbol = new SimpleMarkerSymbol();
markerSymbol.Size = 25;
markerSymbol.Color = Colors.Red;
markerSymbol.Style = SimpleMarkerStyle.X;

SimpleRenderer simpleRenderer = new SimpleRenderer();
simpleRenderer.Symbol = markerSymbol;

featureLayer.Renderer = simpleRenderer;
this.sceneView.Scene.Layers.Add(featureLayer);
```

UniqueValueRenderer

In order to use `UniqueValueRenderer`, you will need to know something about your data. For example, take a look at the following service:

```
https://services.arcgis.com/P3ePLMYs2RVChkJx/ArcGIS/rest/services/
USA_States_Generalized/FeatureServer/0
```

Scroll down the page and note that it has a field named SUB_REGION. This field divides the U.S. into a set of regions. In order to classify this layer based on a unique value, we first need to know the name of the subregions. Scroll down to the bottom of the page and click on the **Query** link. Next, in the **Where** clause, enter 1=1. Then, in **Out Fields**, enter SUB_REGION. Lastly, at the bottom of the page, click on **Query (GET)**. You will now see the SUB_REGION names. There are nine distinct names, which we can now use to build a renderer for this layer. The following lines show the code in its entirety:

```
Uri uri = new
    Uri("https://services.arcgis.com/P3ePLMYs2RVChkJx/ArcGIS/rest/
        services/USA_States_Generalized/FeatureServer/0");

FeatureLayer featureLayer = new FeatureLayer(uri);
```

```
featureLayer.InitializeAsync();

UniqueValueRenderer uniqueValueRenderer = new
    UniqueValueRenderer();
uniqueValueRenderer.Fields.Add("SUB_REGION");

// Define the outline Symbol
SimpleLineSymbol blackSolidOutline = new SimpleLineSymbol();
blackSolidOutline.Color = System.Windows.Media.Colors.Black;
blackSolidOutline.Style = SimpleLineStyle.Solid;
blackSolidOutline.Width = 1;

// Group #1
UniqueValueInfo uiqueValueInfo1 = new UniqueValueInfo();
uiqueValueInfo1.Values.Add("Pacific");
SimpleFillSymbol smpleFillSymbol1 = new SimpleFillSymbol();
smpleFillSymbol1.Color = System.Windows.Media.Colors.Yellow;
smpleFillSymbol1.Outline = blackSolidOutline;
smpleFillSymbol1.Style = SimpleFillStyle.Solid;
uiqueValueInfo1.Symbol = smpleFillSymbol1;
uniqueValueRenderer.Infos.Add(uiqueValueInfo1);

// Group #2
UniqueValueInfo uniqueValueInfo2 = new UniqueValueInfo();
uniqueValueInfo2.Values.Add("West North Central");
SimpleFillSymbol smpleFillSymbol2 = new SimpleFillSymbol();
smpleFillSymbol2.Color = System.Windows.Media.Colors.Blue;
smpleFillSymbol2.Outline = blackSolidOutline;
smpleFillSymbol2.Style = SimpleFillStyle.Solid;
uniqueValueInfo2.Symbol = smpleFillSymbol2;
uniqueValueRenderer.Infos.Add(uniqueValueInfo2);

// Group #3
UniqueValueInfo uniqueValueInfo3 = new UniqueValueInfo();
uniqueValueInfo3.Values.Add("West South Central");
SimpleFillSymbol smpleFillSymbol3 = new SimpleFillSymbol();
smpleFillSymbol3.Color = System.Windows.Media.Colors.Red;
smpleFillSymbol3.Outline = blackSolidOutline;
smpleFillSymbol3.Style = SimpleFillStyle.Solid;
uniqueValueInfo3.Symbol = smpleFillSymbol3;
uniqueValueRenderer.Infos.Add(uniqueValueInfo3);

// Group #4
UniqueValueInfo uniqueValueInfo4 = new UniqueValueInfo();
```

```
uniqueValueInfo4.Values.Add("East North Central");
SimpleFillSymbol smpleFillSymbol4 = new SimpleFillSymbol();
smpleFillSymbol4.Color = System.Windows.Media.Colors.White;
smpleFillSymbol4.Outline = blackSolidOutline;
smpleFillSymbol4.Style = SimpleFillStyle.Solid;
uniqueValueInfo4.Symbol = smpleFillSymbol4;
uniqueValueRenderer.Infos.Add(uniqueValueInfo4);

// Group #5
UniqueValueInfo uniqueValueInfo5 = new UniqueValueInfo();
uniqueValueInfo5.Values.Add("Mountain");
SimpleFillSymbol smpleFillSymbol5 = new SimpleFillSymbol();
smpleFillSymbol5.Color = System.Windows.Media.Colors.Brown;
smpleFillSymbol5.Outline = blackSolidOutline;
smpleFillSymbol5.Style = SimpleFillStyle.Solid;
uniqueValueInfo5.Symbol = smpleFillSymbol5;
uniqueValueRenderer.Infos.Add(uniqueValueInfo5);

// Group #6
UniqueValueInfo uniqueValueInfo6 = new UniqueValueInfo();
uniqueValueInfo6.Values.Add("New England");
SimpleFillSymbol smpleFillSymbol6 = new SimpleFillSymbol();
smpleFillSymbol6.Color = System.Windows.Media.Colors.Cyan;
smpleFillSymbol6.Outline = blackSolidOutline;
smpleFillSymbol6.Style = SimpleFillStyle.Solid;
uniqueValueInfo6.Symbol = smpleFillSymbol6;
uniqueValueRenderer.Infos.Add(uniqueValueInfo6);

// Group #7
UniqueValueInfo uniqueValueInfo7 = new UniqueValueInfo();
uniqueValueInfo7.Values.Add("East South Central");
SimpleFillSymbol smpleFillSymbol7 = new SimpleFillSymbol();
smpleFillSymbol7.Color = System.Windows.Media.Colors.Green;
smpleFillSymbol7.Outline = blackSolidOutline;
smpleFillSymbol7.Style = SimpleFillStyle.Solid;
uniqueValueInfo7.Symbol = smpleFillSymbol7;
uniqueValueRenderer.Infos.Add(uniqueValueInfo7);

// Group #8
UniqueValueInfo uniqueValueInfo8 = new UniqueValueInfo();
uniqueValueInfo8.Values.Add("Middle Atlantic");
SimpleFillSymbol smpleFillSymbol8 = new SimpleFillSymbol();
smpleFillSymbol8.Color = System.Windows.Media.Colors.Gray;
smpleFillSymbol8.Outline = blackSolidOutline;
smpleFillSymbol8.Style = SimpleFillStyle.Solid;
```

```
uniqueValueInfo8.Symbol = smpleFillSymbol8;
uniqueValueRenderer.Infos.Add(uniqueValueInfo8);

// Group #9
UniqueValueInfo uniqueValueInfo9 = new UniqueValueInfo();
uniqueValueInfo9.Values.Add("South Atlantic");
SimpleFillSymbol simpleFillSymbol9 = new SimpleFillSymbol();
simpleFillSymbol9.Color = System.Windows.Media.Colors.Orange;
simpleFillSymbol9.Outline = blackSolidOutline;
simpleFillSymbol9.Style = SimpleFillStyle.Solid;
uniqueValueInfo9.Symbol = simpleFillSymbol9;
uniqueValueRenderer.Infos.Add(uniqueValueInfo9);

featureLayer.Renderer = uniqueValueRenderer;
this.sceneView.Scene.Layers.Add(featureLayer);
```

In this code sample, a `FeatureLayer` class was created and the `UniqueValueRenderer` class was instantiated with the field name. After that, a `SimpleLineSymbol` subclass was created to give the states a solid black outline. Next, each region was symbolized using a `UniqueValueInfo` class by adding value names to it. There are nine unique regions in total. The `UniqueValueInfo` class allows us to assign a `SimpleFillSymbol` class, and then add that to the `Infos` collection of `UniqueValueRenderer`. The last step was then just to apply the renderer to the layer, and then add the layer to `SceneView`.

ClassBreaksRenderer

The `ClassBreaksRenderer` class renderer allows you to render data based on numerical data groups. For example, if you are trying to render the states layer based on income, you would create a `ClassBreakInfo` class for each range of the values. For example, to create a `ClassBreakInfo` class for a range of $50,000 to $100,000, you'd do something like this:

```
ClassBreakInfo classBreakInfo = new ClassBreakInfo();
classBreakInfo.Minimum = 50000;
classBreakInfo.Maximum = 100000;
SimpleFillSymbol simpleFillSymbol2 = new.SimpleFillSymbol();
simpleFillSymbol2.Color = System.Windows.Media.Colors.Green;
simpleFillSymbol2.Outline = blackSolidOutline;
simpleFillSymbol2.Style = SimpleFillStyle.Solid;
classBreakInfo.Symbol = simpleFillSymbol2;
classBreaksRenderer.Infos.Add(classBreakInfo);
```

TemporalRenderer

The last type of renderer is `TemporalRenderer`, which allows you display data from time-enabled layers. There are three kinds of `TemporalRenderer`: `LatestObservationRenderer`, `ObservationRenderer`, and `TackRenderer`.

Military symbols

ArcGIS is ideal for military applications because it provides support for military symbols. Located in `Esri.ArcGISRuntime.Symbology.Specialized`, these classes allow you to render both the **Department of Defense (DoD)** MIL-STD-2525C and **North Atlantic Treaty Organization (NATO)** APP-6B symbols. These symbols are used for situation awareness and command and control, which support secure communications. Troops can send messages by either voice or data transmission over a network using a message, which can then be displayed on a map.

Here's an example of adding `MessageLayer`, and then sending a message using a unique **Symbol ID Code (SIC)**:

```
private async void DisplayMilitaryMessage()
{
    MessageLayer messageLayer = new
        MessageLayer(SymbolDictionaryType.Mil2525c);
    messageLayer.ID = "Military Messages";
    await messageLayer.InitializeAsync();
    this.mapView.Map.Layers.Add(messageLayer);

    // create a dictionary to hold message properties
    var messageProps = new Dictionary<string, string>();

    // define message properties (property name, value)
    messageProps.Add("_type", "position_report");
    messageProps.Add("_action", "update");
    messageProps.Add("_id",
        "e0ba16ac-cddd-4595-845e-32ca4177d080");
    messageProps.Add("_control_points", "71.753904, 36.588076");
    messageProps.Add("_wkid", "4326");
    messageProps.Add("sic", "SHGP----------");
    messageProps.Add("uniquedesignation", "TARGET2");

    // create the message using the properties
    var message = new Message(messageProps);
    // make sure it's loaded before using it
    await  this.mapView.LayersLoadedAsync(new[] { messageLayer });
    bool success = messageLayer.ProcessMessage(message);
}
```

The resulting symbol would look something like this:

As you can see in the code, the message has to be in a certain format, which includes the `_type`, `_action`, `_id`, `_control_points`, `_wkid`, `sic`, and `uniquedesignation` formats. The `_type` format is a position report, the `_action` format is an update, the `_id` format is a GUID, the `_control_points` format is the location, the spatial reference (`_wkid`) is latitude/longitude, the SIC is `"SHGP-----------"`, which is a set of codes for the symbol, and lastly, the `uniquedesignation` format is a unique name for the message. In this example, it's a generic target. For more information about the SIC, see `http://www.dtic.mil/doctrine//doctrine/other/ms_2525c.pdf`. Once the message properties are created, they are passed into a message, which is then processed by the message processor. In a production-level app, these messages would come in at random times, individually or in bulk, and then be processed. The primary benefit of using the message processor is that it can process messages very quickly, which means it can display lots of messages on the map or scene. As a result, it is ideal for situation awareness and command and control applications.

Limitations

In the current version of ArcGIS Runtime, multipatch geometry is not supported. This kind of geometry is available in ArcGIS Desktop and ArcGIS Engine via `ArcObjects`. With multipatch geometry, it is possible to virtually create any shape, such as the clearance cones around a missile's trajectory, by using triangle strips, triangle fans, triangles, and rings. Here's an example:

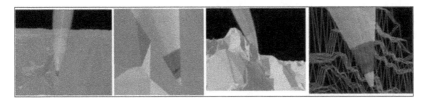

This kind of capability will eventually make it to ArcGIS Runtime as the 3D capabilities are expanded. For now, you can still create multisided shapes with the example code that follows, and apply `PictureFillSymbol` to them as a form of texture if needed:

```
// outline
SimpleLineSymbol sls = new SimpleLineSymbol()
{
    Color = Colors.Black, Style = SimpleLineStyle.Solid, Width = 2
};

// fill symbol
SimpleFillSymbol sfs = new SimpleFillSymbol()
{
    Color = Colors.Red,
    Style = SimpleFillStyle.Null,
    Outline = sls
};

// renderer
SimpleRenderer sr = new SimpleRenderer();
sr.Symbol = sfs;

// Graphics Overlap
GraphicsOverlay graphicsOverlay = new GraphicsOverlay()
{
    RenderingMode = GraphicsRenderingMode.Dynamic,
    Renderer = sr,
    SceneProperties = new LayerSceneProperties()
    {
        SurfacePlacement = SurfacePlacement.Relative
    }
};

// bottom of box
List<MapPoint> bottomList = new List<MapPoint>();
bottomList.Add(new MapPoint(71.75, 36.5, 10000,
    SpatialReferences.Wgs84));
bottomList.Add(new MapPoint(71.75, 37.5, 10000,
    SpatialReferences.Wgs84));
bottomList.Add(new MapPoint(72.75, 37.5, 10000,
    SpatialReferences.Wgs84));
bottomList.Add(new MapPoint(72.75, 36.5, 10000,
    SpatialReferences.Wgs84));
```

```
// West Side of box
List<MapPoint> westSideList = new List<MapPoint>();
westSideList.Add(new MapPoint(71.75, 36.5, 10000,
    SpatialReferences.Wgs84));
westSideList.Add(new MapPoint(71.75, 36.5, 60000,
    SpatialReferences.Wgs84));
westSideList.Add(new MapPoint(72.75, 36.5, 60000,
    SpatialReferences.Wgs84));
westSideList.Add(new MapPoint(72.75, 36.5, 10000,
    SpatialReferences.Wgs84));

// West Side of box
List<MapPoint> eastSideList = new List<MapPoint>();
eastSideList.Add(new MapPoint(72.75, 36.5, 10000,
    SpatialReferences.Wgs84));
eastSideList.Add(new MapPoint(72.75, 36.5, 60000,
    SpatialReferences.Wgs84));
eastSideList.Add(new MapPoint(72.75, 37.5, 60000,
    SpatialReferences.Wgs84));
eastSideList.Add(new MapPoint(72.75, 37.5, 10000,
    SpatialReferences.Wgs84));

// South Side of box
List<MapPoint> southSideList = new List<MapPoint>();
southSideList.Add(new MapPoint(71.75, 36.5, 10000,
    SpatialReferences.Wgs84));
southSideList.Add(new MapPoint(71.75, 36.5, 60000,
    SpatialReferences.Wgs84));
southSideList.Add(new MapPoint(71.75, 37.5, 60000,
    SpatialReferences.Wgs84));
southSideList.Add(new MapPoint(71.75, 37.5, 10000,
    SpatialReferences.Wgs84));

// North Side of box
List<MapPoint> northSideList = new List<MapPoint>();
northSideList.Add(new MapPoint(71.75, 37.5, 10000,
    SpatialReferences.Wgs84));
northSideList.Add(new MapPoint(71.75, 37.5, 60000,
    SpatialReferences.Wgs84));
northSideList.Add(new MapPoint(72.75, 37.5, 60000,
    SpatialReferences.Wgs84));
northSideList.Add(new MapPoint(72.75, 37.5, 10000,
    SpatialReferences.Wgs84));

// Top side of box
List<MapPoint> topList = new List<MapPoint>();
```

```
topList.Add(new MapPoint(71.75, 36.5, 60000,
    SpatialReferences.Wgs84));
topList.Add(new MapPoint(71.75, 37.5, 60000,
    SpatialReferences.Wgs84));
topList.Add(new MapPoint(72.75, 37.5, 60000,
    SpatialReferences.Wgs84));
topList.Add(new MapPoint(72.75, 36.5, 60000,
    SpatialReferences.Wgs84));

// create polygons of box
Polygon bottomPolygon= new Polygon(bottomList);
Polygon topPolygon = new Polygon(topList);
Polygon westSidePolygon = new Polygon(westSideList);
Polygon southSidePolygon = new Polygon(southSideList);
Polygon northSidePolygon = new Polygon(northSideList);
Polygon eastSidePolygon = new Polygon(eastSideList);

// create graphics of box
Graphic bottomGraphic = new Graphic(bottomPolygon);
Graphic topGraphic = new Graphic(topPolygon);
Graphic westSideGraphic = new Graphic(westSidePolygon);
Graphic southSideGraphic = new Graphic(southSidePolygon);
Graphic northSideGraphic = new Graphic(northSidePolygon);
Graphic eastSideGraphic = new Graphic(eastSidePolygon);

// add graphics to overlap
graphicsOverlay.Graphics.Add(bottomGraphic);
graphicsOverlay.Graphics.Add(topGraphic);
graphicsOverlay.Graphics.Add(westSideGraphic);
graphicsOverlay.Graphics.Add(southSideGraphic);
graphicsOverlay.Graphics.Add(northSideGraphic);
graphicsOverlay.Graphics.Add(eastSideGraphic);

// add overlay
this.sceneView.GraphicsOverlays.Add(graphicsOverlay);
```

The resulting box will look like the following screenshot, and could represent an airspace deconfliction area:

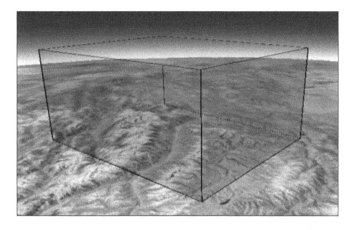

Summary

In this chapter, we've covered geometry and symbology. We've discussed how to assign geometry to the subclasses of Feature, we've gone over how to construct immutable geometric objects such as points, polyline, and polygons, as well as mutable geometric objects, units of measure, the geometry engine, and symbology, which included simple symbols, scene symbols, military symbols, renderers, and even some of the current limitations of geometry at this release of ArcGIS Runtime.

In the next chapter, we will expand on different ways to display information on the map or scene by interacting with the layers and showing information using a variety of techniques.

6
Displaying Information

So far, we've created maps and scenes, added layers, learned about geometry and symbology, and so on, but now we need to learn about displaying and interacting with the map or scene. Adding information such as graphics has already been discussed, but what if you want to see the mouse's coordinates in latitude/longitude? What if you have lots of layers and want to have overlays (map tips) for each one of them that are customized for the layer? How do you show a legend, so users can discern what is on the map? Lastly, how do you show your location on the map or scene? All of these are questions that production-level apps need to address so that users can quickly understand what is being presented and where they are. In this chapter, we will discuss the following topics:

- MVVM mouse events
- Labeling
- Legend
- Scale bar
- Overlays
- Showing locations

MVVM mouse events

Being able to interact with the layers in the map is a very important capability that your users will expect. Otherwise, the map is just a pretty picture. To provide this capability in a non-MVVM app would simply require adding a click event to the code-behind file of `MainWindow.cs`. Another feature that your user might expect is the ability to see coordinates on the map as they move the mouse. But how do we accomplish this with MVVM in such a way that we can move our code from `MainWindow.cs` to another place?

Well, you already know the answer. We create a `UserControl` task with `ViewModel` and have it display the coordinates in the XAML code of `UserControl`. Let's implement this:

1. Create a new ArcGIS Runtime project. In this example, we're going to name it `Chapter6`. Copy the following folders: `Behavior`, `Models`, `Services`, and `ViewModels`. Install MVVM Light and `Json.Net`, and add a reference to `System.Windows.Interactivity` (4.5). When you install MVVM Light, a `ViewModel` directory will be created. Delete it. Update `App.xaml` with a reference to the `Chapter6.ViewModels` folder. Update all other references and the `using` statements. Copy the XAML code from `MainWindow.xaml` in `Chapter3a` to the `MainWindows.xaml` file in your new app. Make sure the app builds.

2. Create a new folder and name it `UserControls`. Add a new `UserControl` (WPF) item to this app. Name it `CoordinateDisplayUserControl.xaml`.

3. Create a new `ViewModel` class in the `ViewModels` folder and name it `CoordinateDisplayViewModel.cs`.

4. Add the following `using` statements to `CoordinateDisplayViewModel.cs`:

```
using GalaSoft.MvvmLight.Messaging;

using Esri.ArcGISRuntime.Controls;
using Esri.ArcGISRuntime.Geometry;
using System;
```

5. In the constructor, add the following lines. When you create the `MouseMove` event handler, press *Tab* twice to create it:

```
Messenger.Default.Register<Esri.ArcGISRuntime.Controls.
    MapView>(this, (mapView) =>
{
    this.mapView = mapView;

    this.mapView.MouseMove += mapView_MouseMove;
});
```

6. In the `MouseMove` event handler (`mapView_mouseMove`), add the following lines:

```
if (this.mapView.Extent == null)
    return;
System.Windows.Point screenPoint =
    e.GetPosition(this.mapView);
MapPoint mapPoint =
    this.mapView.ScreenToLocation(screenPoint);
if (this.mapView.WrapAround)
```

```
mapPoint =
    GeometryEngine.NormalizeCentralMeridian(mapPoint)
        as MapPoint;

this.Latitude = Math.Round(mapPoint.Y, 4);
this.Longitude = Math.Round(mapPoint.X, 4);
```

As you can see, we are taking the `Screen` coordinate and converting it to `MapPoint` using `ScreenToLocation`. If `WrapAround` is turned on, we use the `GeometryEngine` class to normalize the `MapPoint` class. Finally, we get the coordinates and store them in two properties so that we can bind to them.

7. Add the following two properties to this class:

```
public double Latitude
{
    set
    {
        this.latitude = value;
        RaisePropertyChanged("Latitude");
    }
    get { return this.latitude; }
}
public double Longitude
{
    set
    {
        this.longitude = value;
        RaisePropertyChanged("Longitude");
    }
    get { return this.longitude; }
}
```

8. Add the new `ViewModel` class to the locator (`ViewModelLocator.cs`):

```
SimpleIoc.Default.Register<CoordinateDisplayViewModel>();
```

9. Open up `CoordinateDisplayUserContro.xaml` and enter the following code:

```
<UserControl
    x:Class="Chapter6.UserControls.
        CoordinateDisplayUserControl"
    xmlns="http://schemas.microsoft.com/winfx/2006/xaml/
        presentation"
    xmlns:x="http://schemas.microsoft.com/winfx/2006/xaml"
    xmlns:mc="http://schemas.openxmlformats.org/
        markup-compatibility/2006"
```

```
xmlns:d="http://schemas.microsoft.com/expression/blend/
    2008"
xmlns:vm="clr-namespace:Chapter6.ViewModels"
mc:Ignorable="d"
d:DesignHeight="300" d:DesignWidth="500">

<UserControl.Resources>
    <vm:CoordinateDisplayViewModel x:Key="Locator"
        d:IsDataSource="True"/>
</UserControl.Resources>

<Grid DataContext="{Binding CoordinateDisplayViewModel,
    Source={StaticResource Locator}}">
    <StackPanel Orientation="Horizontal">
        <Border  Background="Transparent"
            BorderBrush="Black"
                BorderThickness="1"
            HorizontalAlignment="Left"
                VerticalAlignment="Top"
            Margin="30" Padding="20">
            <Border.Effect>
                <DropShadowEffect/>
            </Border.Effect>
            <StackPanel >
                <StackPanel>
                    <StackPanel
                        Orientation="Horizontal">
                        <TextBlock x:Name="Latitude"
                            Foreground="White"
                            HorizontalAlignment="Left"
                            VerticalAlignment="Center"
                            Text="Latitude: "
                                Width="80"
                                TextWrapping="Wrap"
                                FontWeight="Bold" />
                        <TextBlock
                            x:Name="LatitudeValue"
                            Foreground="White"
                            HorizontalAlignment="Left"
                            VerticalAlignment="Center"
                            TextAlignment="Right"
                            Text="{Binding Latitude,
                                Mode=TwoWay,
                            UpdateSourceTrigger=
                                PropertyChanged,
```

```
                    Source={StaticResource
                        Locator}}"
                    Width="80"
                        FontWeight="Bold" />
            </StackPanel>
            <StackPanel
                Orientation="Horizontal">
                <TextBlock x:Name="Longitude"
                    Foreground="White"
                    HorizontalAlignment="Left"
                    VerticalAlignment="Center"
                    Text="Longitude: "
                    TextWrapping="Wrap"
                    FontWeight="Bold" />
                <TextBlock
                    x:Name="LongitudeValue"
                    Foreground="White"
                    HorizontalAlignment="Left"
                    VerticalAlignment="Bottom"
                    TextAlignment="Right"
                    Text="{Binding Longitude,
                        Mode=TwoWay,
                    UpdateSourceTrigger=
                        PropertyChanged,
                    Source={StaticResource
                        Locator}}"
                    Width="95"
                        FontWeight="Bold" />
            </StackPanel>
            </StackPanel>
            </StackPanel>
        </Border>
        </StackPanel>
    </Grid>
</UserControl>
```

Note that the resource of `UserControl` has been set to the locator and that the `Grid` tag has been set to `CoordinateDisplayViewModel`. Other than that, the two `TextBlock` properties are binding to the `ViewModel` class' properties (`Latitude` and `Longitude`).

10. Add the new user control to the `MainWindow.xaml` file by first adding a `using` statement to `MainWindow.xaml` like this:

```
xmlns:uc="clr-namespace:Chapter6.UserControls"
```

11. Then, add the user control after the closing `</MapView>` tag like this:

```
<uc:CoordinateDisplayUserControl
    VerticalAlignment="Top">
        </uc:CoordinateDisplayUserControl>
```

12. Run the app. As you move the mouse over the map, you will see the coordinates update based on the mouse's position, as we can see here:

Coordinates based on the mouse's position

If you recall, in *Chapter 1, Introduction to ArcGIS Runtime*, we did something similar with the map scale but that was with the code-behind file. This time, we did it with a more MVVM-friendly approach. Of course, we could refactor this code to pass off the coordinates to a factory service, but for the sake of brevity we left this out.

While we used a `UserControl` task with a `ViewModel` class to solve this issue, we could have used other approaches too. We could have used `Behavior`, `TargetedTriggerAction`, or `ChangePropertyAction`. Look them up in Microsoft's documentation, give them a try, and see which one you prefer.

Labeling

Looking at a bunch of points, a polyline of polygons on the map provides the location of the content, but it doesn't provide enough context about these geometries, even if they have beautiful symbology. Sometimes, it's necessary to label the features with text. Labeling is a cartographic term that means placing text on or near a feature in order to present contextual information to the user. It's possible to use a `TextSymbol` class and iterate over every feature or graphic, but that would be slow. Fortunately, ArcGIS Runtime comes with a labeling engine that allows you to label each feature or graphic with text in an efficient manner. In fact, with a little code you can label `FeatureLayer` with hundreds and thousands of features. It also possible to label a `GraphicsLayer` or `GraphicOverlay` class.

To label a layer, there are few classes in `Esri.ArcGISRuntime.Layers` that we need to become familiar with, so let's discuss them, and then see some examples of how to label and the options that we have. See here:

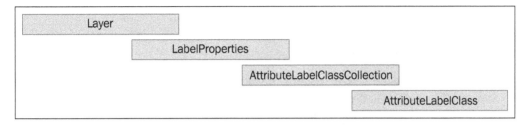

The `LabelProperties` class allows you to enable labeling on a layer. Just set `IsEnabled` to `True`. The `FeatureLayer`, `GraphicsLayer`, and `GraphicOverlay` classes have a property named `Labeling`, which is where you set the `LabelProperties` class to.

When you set up labeling, you do so with `AttributeLabelClass`. But you can set up labeling with multiple fields in a `FeatureLayer` class, for example. To accomplish this, you would create a collection of `AttributeLabelClass`, and then add them to a `AttributeLabelClassCollection` class. Once you have the `AttributeLabelClassCollection` class filled with `AttributeLabelClass`, you can then apply it to a `LabelProperties` class, which is then set to the `Labeling` property of the layer. Here is an example of setting up labeling with a dynamic map service:

```
<esri:FeatureLayer ID="Cities" >
    <esri:ServiceFeatureTable
        ServiceUri="http://sampleserver5.arcgisonline.com/arcgis/
            rest/services/USA/MapServer/0"/>
    <esri:FeatureLayer.Labeling>
        <esri:LabelProperties IsEnabled="True">
            <esri:AttributeLabelClass WhereClause="capital = 'N'"
                TextExpression="[areaname]"
            LabelPlacement="PointAboveCenter">
                <esri:TextSymbol Color="OldLace">
                    <esri:SymbolFont FontFamily="Segoe"
                        FontSize="14"
                    FontWeight="Bold"/>
                </esri:TextSymbol>
            </esri:AttributeLabelClass>
            <esri:AttributeLabelClass WhereClause="capital = 'Y'"
                TextExpression="[areaname]"
            LabelPlacement="PointAboveCenter">
                <esri:TextSymbol Color="white">
                    <esri:SymbolFont FontFamily="Segoe"
                        FontSize="18"
```

```
                    FontWeight="Bold"/>
                </esri:TextSymbol>
            </esri:AttributeLabelClass>
        </esri:LabelProperties>
    </esri:FeatureLayer.Labeling>
</esri:FeatureLayer>
```

And here is an example using code:

```
// Create a new Uri that points to a FeatureLayer.
System.Uri myUri = new
    System.Uri("http://sampleserver6.arcgisonline.com/arcgis/rest/
        services/USA/MapServer/1");

// Create a new ServiceFeatureTable and set it's ServiceUri and
    Where
// clause Properties.
ServiceFeatureTable serviceFeatureTable = new
    ServiceFeatureTable();
serviceFeatureTable.ServiceUri = myUri.ToString();

// Returns all the fields in the FeatureLayer.
// NOTE: If the Attribute Field that needs to be labeled via the
// AttributeLabelClass.TextExpression is not returned, then no
// labeling will occur.
serviceFeatureTable.OutFields =
    Esri.ArcGISRuntime.Tasks.Query.OutFields.All;

// Create a new instance of an FeatureLayer and set it's Id and
    FeatureTable Properties.
FeatureLayer featureLayer = new FeatureLayer();
featureLayer.ID = "Highways";
featureLayer.FeatureTable = serviceFeatureTable;

// Define a SimpleLineSymbol for the highways.
SimpleLineSymbol simpleLineSymbol = new SimpleLineSymbol();
simpleLineSymbol.Color = System.Windows.Media.Colors.Black;
simpleLineSymbol.Style = SimpleLineStyle.Solid;
simpleLineSymbol.Width = 1;

// Create a SimpleRenderer to hold the SimpleLineSymbol.
SimpleRenderer simpleRenderer = new SimpleRenderer();
simpleRenderer.Symbol = simpleLineSymbol;
```

```
// Apply the SimpleRenderer to the FeatureLayer.
featureLayer.Renderer = simpleRenderer;

// Add the FeatureLayer to the Map.
MyMapView.Map.Layers.Add(featureLayer);

// ------------------------------------------------------------
// Creating and adding the Labeling option to the FeatureLayer.
// ------------------------------------------------------------

// Create an AttributeLabelClass. This contains the meat of the
// instructions for doing labeling.
AttributeLabelClass attributeLabelClass = new
    AttributeLabelClass();
attributeLabelClass.DuplicateLabels =
DuplicateLabels.RemoveWithinLabelClass;
attributeLabelClass.IsVisible = true;
attributeLabelClass.IsWordWrapEnabled = false;
attributeLabelClass.LabelPlacement =
    LabelPlacement.LineAboveAlong;
attributeLabelClass.LabelPosition =
LabelPosition.FixedPositionOrRemove;
attributeLabelClass.LabelPriority = LabelPriority.Medium;
attributeLabelClass.MaxScale = 0;
attributeLabelClass.MinScale = 0;
attributeLabelClass.TextExpression = "\"" + "I - " + "\"" + "
    CONCAT [rte_num1]";
attributeLabelClass.WhereClause = "[rte_num1] <> ' '";
attributeLabelClass.WordWrapLength = 999;

// Create a new TextSymbol to define the appearance of the text
    that
// is displayed.
TextSymbol textSymbol = new TextSymbol();
textSymbol.Color = System.Windows.Media.Colors.Black;
textSymbol.BorderLineColor = System.Windows.Media.Colors.Yellow;
textSymbol.BorderLineSize = 3;

// Create a new SymbolFont to define the appearance of the text
    that
// is displayed.
SymbolFont mySymbolFont = new SymbolFont();
mySymbolFont.FontFamily = "Verdana";
mySymbolFont.FontSize = 12;
```

```
mySymbolFont.FontStyle = SymbolFontStyle.Normal;
mySymbolFont.TextDecoration = SymbolTextDecoration.None;
mySymbolFont.FontWeight = SymbolFontWeight.Bold;
textSymbol.Font = mySymbolFont;

// Apply the TextSymbol to the AttributeLabelClass.Symbol
    Property.
// IMPORTANT: This value must be provided or else no labeling will
// occur!
attributeLabelClass.Symbol = textSymbol;

// Create a new AttributeLabelClassCollection to hold one or more
// AttributeLabelClass'es.
AttributeLabelClassCollection attributeLabelClassCollection = new
    AttributeLabelClassCollection();
attributeLabelClassCollection.Add(attributeLabelClass);

// Create a new LabelProperties to hold the
// AttributeLabelClassCollection. Make sure it is enabled (i.e.
// Visible).
LabelProperties labelProperties = new LabelProperties();
labelProperties.IsEnabled = true;
labelProperties.LabelClasses = attributeLabelClassCollection;

// Apply the LabelProperties to the FeatureLayer.Labeling Property.
featureLayer.Labeling = labelProperties;

// Enable labeling on the GraphicsLayer as well.
featureLayer.Labeling.IsEnabled = true;
```

There are several items of interest that need further explanation.

The DuplicateLabels class provides an enumeration that allows you to specify whether to preserve duplicates (PreserveDuplicates), remove duplicates within the label class (RemoveWithinLabelClass), remove a geometry type from all layers (RemoveByGeometryTypeFromAllLayers), and lastly, you can remove all duplicates from all layers (RemoveAllDuplicatesFromAllLayers). The ability to remove duplicates is very important when working with polylines such as roads. If a road is composed of multiple polylines, the option to remove duplicates can reduce the labeling clutter, which improves performance. See here:

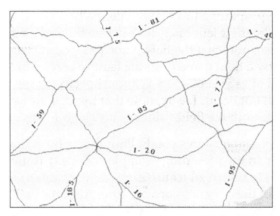

Option to remove duplicates reduces the labeling clutter

The LabelPlacement class provides you with the ability to control how the label is placed around points, polylines, or polygons. The LabelPlacement class is composed of the following three parts:

- The geometry type must match the type of LabelPlacement. For example, PointAboveCenter goes with point geometry. On the other hand, LineAboveBefore goes with polyline geometry.
- For the vertical position, we use Above, Below, or Center.
- For the horizontal position, we use Center, Left, Right (Point), Before, Start, Along, End, After (Line).

Some examples of LabelPlacement are PointAboveCenter, PointAboveRight, LineAboveAlong, LineAboveStart, PolygonAlwaysHorizontal, and so on. This last option, PolygonAlwaysHorizontal, deserves a special note. When the labeling engine tries to label a polygon, it will attempt to place the label relative to the first point in the polygon. If you use PolygonAlwaysHorizontal, it will use the polygon's center of mass.

The LabelPosition property determines the label's positioning behavior. The labeling algorithm attempts to place as many labels as possible, based on the properties specified. If you set LabelPosition to FixedPositionOrRemove as shown in the preceding code, this option will result in more labels being produced but will also reduce performance. There are two other options: FixedPositionWithOverlaps, which will force all labels to be placed while keeping overlapping labels, and RepositionOrRemove, which will attempt to reposition or remove labels that can't be repositioned.

The LabelPriority enumeration determines which label AttributeLabelClass has higher priority. If two or more AttributeLabelClass instances are defined, the one with the highest priority will be displayed with higher priority.

The TextExpression property allows you to define how the label will appear along the feature. In this code, the I letter is placed before the route number (rte_num1), along with a dash. Also, note that the field name (rte_num1) has brackets around it. This is how you specify a field name from the feature service in this case. This works the same for any kind of FeatureLayer. Without specifying the TextExpression property, labeling will not occur. Lastly, note that the CONCAT keyword is used to concatenate the literals with the field values.

The WhereClause expression allows you to limit the features that are labeled using a SQL-like expression. In the preceding example code, any route that has some text is labeled. If you want to return all features, use WhereClause such as '1=1' or 'True='True'. Both of these expressions are true. Also, it's important that you know the data type and you need to make sure, as in the preceding example, the fields you are using in WhereClause are returned via ServiceFeatureTable.OutFields. Otherwise, WhereClause won't know the fields to use when searching.

Finally, you can set the Symbol property of the text using TextSymbol and the minimum and maximum scale at which the labels are drawn. It is always recommended to set the minimum and maximum scale to improve the performance of the labeling engine.

Legend

Most apps have more than one layer. In real apps, there could be five layers or even 20 layers. You could even have the option for the user to have multiple basemaps. Once multiple vector (geometric) layers are added to the map, it can easily become difficult to discern which layers are which, especially in dense urban areas. Because of this, a legend is really helpful and pretty much a standard feature for non-trivial apps. Here is an example of a legend built with the iOS version of the SDK:

Legend built with the ios version

Unfortunately, ArcGIS Runtime does not come with a legend control in `Esri.ArcGISRuntime.Controls`. However, Esri has built a toolkit that comes with some sample legends, which we will now make use of in the next exercise:

1. Add a reference to `Esri.ArcGISRuntime.Toolkit.dll`, which is located in `C:\ArcGISRuntimeBook`.

2. In the project we created earlier in this chapter, we will now add some XAML code to `MainWindow.xaml`. Add the following reference to the XAML code:

```
xmlns:toolkit="clr-namespace:Esri.ArcGISRuntime.Toolkit.
    Controls;assembly=Esri.ArcGISRuntime.Toolkit"
```

3. After the closing `</MapView>` tab, add the following XAML code:

```
<Border  BorderBrush="Black" BorderThickness="1"
    Margin="30"
    Padding="10,20,20,20"
    HorizontalAlignment="Right" VerticalAlignment="Top">
    <Border.Effect>
        <DropShadowEffect />
    </Border.Effect>
    <ScrollViewer VerticalScrollBarVisibility="Hidden">
        <toolkit:Legend Foreground="White"
            Layers="{Binding Map.Layers,
            ElementName=MyMapView}"
                ReverseLayersOrder="False" >
        </toolkit:Legend>
    </ScrollViewer>
</Border>
```

4. Run the app and you will note that you have a legend:

It's not much at this point because we don't have many layers; however, as you add more layers, the legend will show any new FeatureLayer classes. All that this code does is use a component in the toolkit named Legend. The Legend component has a property called Layers, which is bound to the MapView component's Map, which has a Layers property. The layers are also shown in the reverse order in which they were added using the ReverseLayersOrder property.

There are other kinds of legend in the toolkit, which you can explore on your own. In fact, you can download the source code for the toolkit from GitHub at https://github.com/Esri/arcgis-toolkit-dotnet.

Scale bar

While we're at it, let's add a scale bar to our map so that users can get a sense of distances:

1. In MainWindow.xaml, add the following after the legend code:

    ```
    <toolkit:ScaleLine Scale="{Binding ElementName=MyMapView,
        Path=Scale}"
        HorizontalAlignment="Right" VerticalAlignment="Bottom"
            Margin="10" />
    ```

2. Run the app and you will see a scale bar like the following screenshot in the lower-right corner of the map:

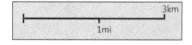

Make sure to zoom in and out and you'll note that it changes, depending on the scale of the map. Once again, ScaleLine has a property called Scale, which binds to the Scale property of MapView. As we discussed in *Chapter 4, From 2D to 3D*, there is no such thing as scale in 3D, so this control won't work in SceneView.

Overlays

In earlier chapters, we discussed overlays in a limited manner. But now, it's time to go a little deeper into the subject. Overlays allow you to add .NET FrameworkElement to a collection called OverlayItemsCollection. For example, the following code can be used to add TextBlock over San Francisco:

```
var tb = new TextBlock();
tb.Foreground = new SolidColorBrush(Colors.Red);
tb.FontWeight = FontWeights.Bold;
```

```
tb.Background = new SolidColorBrush(Colors.White);
tb.Height = 20;
tb.Text = "San Francisco";

var geoPoint = new Esri.ArcGISRuntime.Geometry.MapPoint(-122.44,
    37.8, Esri.ArcGISRuntime.Geometry.SpatialReferences.Wgs84);

this.mapView.Overlays.Items.Add(tb);
    Esri.ArcGISRuntime.Controls.MapView.SetViewOverlayAnchor(tb,
    geoPoint);
```

See here:

Addition of text block

A `TextBlock` component such as this adds a little context to the map, but you can go way beyond that because you can add pretty much anything into the `Overlays` collection. Typically, however, overlays are added to the map to emphasize something, but they should be used sparingly because they consume memory quickly and therefore reduce performance. Regardless, they allow you to also display attribute information for features too, so that a user can click on a feature and see the field values, as in the following screenshot:

In this case, the user has clicked on a parking meter and the ID and street name appear over it.

Now that you understand the basics of overlays, it's time to take a deeper dive into the subject so that we can create them for lots of layers. When adding overlays to your `FeatureLayer` classes, you have a couple of options: static or dynamic XAML.

Static XAML

Static XAML is used to create XAML at design time to add overlays. They are simple to use, as shown here:

```
<esri:MapView.Overlays>
    <esri:OverlayItemsControl>
        <Border x:Name="mapTip" Background="White"
            BorderBrush="Black"
            BorderThickness="2"
            Padding="4" Margin="4"
            HorizontalAlignment="Left"
            VerticalAlignment="Top"
            Visibility="Collapsed"
            esri:ViewBase.ViewOverlayAnchor="{Binding Geometry}">
            <StackPanel>
                <TextBlock Text="{Binding
                    Attributes[facname]}" FontWeight="Bold" />
                <TextBlock Text="{Binding
                    Attributes[capacity],
                    StringFormat='Capacity: {0}'}" />
            </StackPanel>
        </Border>
    </esri:OverlayItemsControl>
```

A `FeatureLayer` tag could then be added, either with XAML or with a code such as this:

```
<esri:FeatureLayer.FeatureTable>
    <esri:ServiceFeatureTable
    ServiceUri="http://sampleserver6.arcgisonline.com/arcgis/rest/
        services/NapervilleShelters/FeatureServer/0"
        OutFields="*" />
</esri:FeatureLayer.FeatureTable>
```

In this code, the overlays are defined with .NET `FrameworkElements` (the `TextBlock` tags inside the `StackPanel` tags, which are in turn inside a `Border` tag). Then, `FeatureLayer` is created. When the user clicks on a feature, it shows a map tip like this:

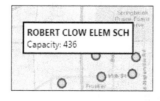

Dynamic XAML

Static XAML is fine with trivial apps; however, if you want to provide the user with the ability to add layers from disk or from the Web, or you have a configuration store that allows users to add, edit, or delete layers, then the app won't know what fields it needs to display, as shown in the preceding example with a hardcoded layer and field names. Another approach to generating XAML is to generate it dynamically at runtime.

Let's modify the app we've been using in this chapter that works with San Francisco data so that you can click on a meter and display the meter ID and street name. First, we'll use static XAML, and then we'll refactor the code to use dynamic XAML:

1. In `MainWindow.xaml`, add the following XAML code after the closing `</Map>` tag:

```
<esri:MapView.Overlays>
    <!-- OverlayItemsControl for the MapView. -->
    <esri:OverlayItemsControl x:Name="overlayItemsControl">
        <Border
            CornerRadius="10"
            BorderBrush="Black"
            Margin="0,0,25,25" Visibility="Hidden"
            BorderThickness="2" Background="#995C90B2" >

            <StackPanel Orientation="Vertical"
                Margin="5,10,18,15">
                <StackPanel Orientation="Horizontal">
                    <TextBlock Text="ID: "
                        FontWeight="Normal"
                        Height="20" Foreground="White"/>
                    <TextBlock Text="{Binding [POST_ID]}"
                        FontWeight="Normal"
                        Foreground="White"  Height="20"/>
                </StackPanel>

                <StackPanel Orientation="Horizontal">
                    <TextBlock Text="Street: "
                        FontWeight="Normal"
                        Height="20" Foreground="White"/>
                    <TextBlock Text="{Binding
                        [STREETNAME]}"
                        FontWeight="Normal"
                        Foreground="White"  Height="20"/>
                </StackPanel>
            </StackPanel>
```

```
        </Border>
     </esri:OverlayItemsControl>
  </esri:MapView.Overlays>
```

2. In `MapViewBehavior.cs`, add a `using` statement to `Esri.ArcGISRuntime.`
    `Layers`, and then add a new `MapViewTapped` event handler to the
    `OnAttached` method, like this:

```
AssociatedObject.MapViewTapped +=
    AssociatedObject_MapViewTapped;
```

3. In `AssociatedObject_MapViewTapped`, add the following code:

```
MapView mapView = sender as MapView;
OverlayItemsControl overlays = mapView.Overlays;
OverlayItemsCollection collection = overlays.Items;
System.Windows.Controls.Border overlayControl =
    (System.Windows.Controls.Border)collection[0];

// get the location tapped on the map
var mapPoint = e.Location;

// Get possible features and if none found, move to next
    layer
FeatureLayer featureLayer =
    (FeatureLayer)mapView.Map.Layers[1];

// Get possible features and if none found, move to next
    layer
var foundFeatures = await
    featureLayer.HitTestAsync(mapView, new Rect(e.Position,
        new Size(10, 10)), 1);

if (foundFeatures.Count() == 0)
            return;

var feature = await
    featureLayer.FeatureTable.QueryAsync(foundFeatures[0]);

overlayControl.DataContext = feature.Attributes;

Esri.ArcGISRuntime.Controls.MapView.SetViewOverlayAnchor(
    overlayControl, mapPoint);
overlayControl.Visibility = Visibility.Visible;;
```

4. Run the app, and then click on any parking meter. You will see something like the following screenshot:

What we've done here is follow the principles of MVVM by placing the `MapViewTapped` event handler in the custom `MapView` behavior, while at the same time, we've added useful information to the map. The user can now click on a parking meter and find out details. We only added a couple of fields to illustrate the concept, but now let's go over the tapped event handler.

When the user clicks on the map, `MapViewTapped` is fired. As we've previously seen, `MapView` comes along and we simply cast it from sender to `MapView`. Then, we get the overlays in `Map` that were defined in the XAML code. After this, we get the collection that makes up the overlay, which in this case is a single border. From there, we get the location that the user clicks on and cast it to `MapPoint`. Next, we get the second layer in the map, and then perform `HitTest`, which returns the features found in the area we clicked. The `HitTestAsync` method uses a `Rect` parameter as a tolerance parameter. After the feature is found, we query `FeatureLayer` asynchronously and return a single feature. We only want one feature so that we can bind its attributes to the overlay, which in this case is a border with .NET `FrameworkElements` inside it. The feature's attributes are bound to the border's data context. After that, we set the overlays anchor position, which is `MapPoint` where we clicked. Finally, we set the border's visibility to `Visible` so that it shows itself.

Now that we've implemented map tips using static XAML, let's refactor this code to use dynamic XAML so that the overlays are created at runtime:

1. In the project that we've been working on, there is a method in `MainViewModel.cs` that looks like this:

```
private async void CreateFeatureLayers()
{
    var gdb = await Geodatabase.OpenAsync(this.GDB);

    Envelope extent = null;
    foreach (var table in gdb.FeatureTables)
    {
        var flayer = new FeatureLayer()
        {
```

```
                ID = table.Name,
                DisplayName = "Parking Meters",
                FeatureTable = table
        };

        if
            (!Geometry.IsNullOrEmpty
                (table.ServiceInfo.Extent))
        {
            if (Geometry.IsNullOrEmpty(extent))
                extent = table.ServiceInfo.Extent;
            else
                extent =
                    extent.Union(table.ServiceInfo.Extent);
        }

        this.mapView.Map.Layers.Add(flayer);
    }

    await this.mapView.SetViewAsync(extent.Expand(1.10));
}
```

2. Let's add the following code after the `FeatureLayer` class, or you can just copy it from the `Chapter6` project that came with this book:

```
OverlayItemsControl overlays = this.mapView.Overlays;
OverlayItemsCollection collection = overlays.Items;

System.Text.StringBuilder sb = new
    System.Text.StringBuilder(@"<Border CornerRadius=""10""
    xmlns=""http://schemas.microsoft.com/winfx/2006/xaml/
        presentation""
    BorderBrush=""Black""
    Margin=""0,0,25,25"" Visibility=""Hidden""
    BorderThickness=""2"" Background=""#995C90B2"" >");
sb.Append(@" <StackPanel Orientation=""Vertical""
    Margin=""5,10,18,15"">");
sb.Append(@" <StackPanel Orientation=""Horizontal"">");

sb.Append(@" <TextBlock Text=""ID: "" FontWeight=""Normal""
    Height=""20"" Foreground=""White""/>");
sb.Append(@"  <TextBlock Text=""{Binding [POST_ID]}""
    FontWeight=""Normal"" ");
sb.Append(@" Foreground=""White""  Height=""20""/>");
sb.Append(@"   </StackPanel>");
sb.Append(@" <StackPanel Orientation=""Horizontal"">");
```

```
sb.Append(@"   <TextBlock Text=""Street: ""
    FontWeight=""Normal"" Height=""20"" Foreground=""White""/>");
sb.Append(@"   <TextBlock Text=""{Binding [STREETNAME]}""
    FontWeight=""Normal"" ");
sb.Append(@"   Foreground=""White""   Height=""20""/>");
sb.Append(@"   </StackPanel>");
sb.Append(@"   </StackPanel>");
sb.Append(@"   </Border>");


System.Windows.FrameworkElement element =
    (System.Windows.FrameworkElement)System.Windows.Markup.
        XamlReader.Parse(sb.ToString());


collection.Add(element);
```

3. In `MainWindow.xaml`, comment out the overlays code. This includes everything from `<esri:MapView.Overlays>` to `</esri:MapView.Overlays>`.

4. Run the app. It should work just like it did earlier. Click on a parking meter and it shows a map tip.

In the code, we got a reference to the overlays, which at first will have nothing in it. Then, we get the items in the overlay and pass it to `OverlayItemsCollection`. After that, we create `StringBuilder`, and then populate it with the exact same XAML code we used earlier, except this time we added a reference to the XAML namespace. This allows our overlays to understand that this is an XAML markup; otherwise, it's just a string of XML. The other difference is that we had to escape all of the string representations of the XAML code by adding double quotes where there were single quotes. After the XAML code was created, we converted `StringBuilder` to a string and passed it to `XamlReader`, which translated our string to a .NET `FrameworkElement`. Lastly, we added the element to `OverlayItemsCollection`.

What's the difference? Well, instead of defining the XAML code in `MainWindow. xaml`, we defined it in code, but more importantly, this XAML code was created at runtime instead of design time. But, think bigger about what you could do with this. We could have read this from a configuration file instead of hardcoding the field names in the XAML code. This way, our XAML code would be truly dynamic. We could even change the XAML code to be not only based on a configuration file, but also based on the type of user or the scale, or anything. There are many possibilities with this approach, and hopefully, this will give you some ideas to make your apps even more interesting.

Showing locations

If you are making a mapping app these days without providing the user the option to show their location, you've kind of missed the 21st century boat, so to speak. For years now, this has been a standard feature of Google and Apple's mapping technology. ArcGIS Runtime comes with everything you will need to display your current location and more:

Displays the current location

Let's look at the classes that you will need to use. First, `LocationDisplay` is a property of `MapView`. It contains everything you need to show the current location. Here's an example of how it is done:

```
<Window.Resources>
    <esri:SimpleMarkerSymbol x:Key="NotMovingMarkerSym"
        Color="Red"
        Style="X" Size="16"/>
    <esri:SimpleMarkerSymbol x:Key="MovingMarkerSym" Color="Blue"
        Style="Circle" Size="12"/>
</Window.Resources>
<Grid>
    <esri:MapView x:Name="MyMapView">
        <esri:MapView.LocationDisplay>
            <esri:LocationDisplay
                AutoPanMode="Navigation"
                DefaultSymbol="{StaticResource
                    NotMovingMarkerSym}"
                CourseSymbol="{StaticResource MovingMarkerSym}"
                IsEnabled="True" IsPingAnimationEnabled="True">
            </esri:LocationDisplay>
        </esri:MapView.LocationDisplay>
    <esri:Map>
        <esri:ArcGISTiledMapServiceLayer ID="Basemap"
            ServiceUri="http://services.arcgisonline.com/arcgis/
                rest/services/World_Street_Map/MapServer"/>
```

```
        </esri:Map>
      </esri:MapView>
  </Grid>
```

As you can see, the `LocationDisplay` object is instantiated in this XAML code, and then the properties of `LocationDisplay` are set. These properties include `AutoPanMode`, `DefaultSymbol`, `IsEnabled`, and `IsPingAnimationEnabled`. The most important property is `IsEnabled`. This property tells the `LocationDisplay` object to start showing your location. However, in order for this to work on a desktop or laptop, you will need to install an external GPS. If your mobile device has a built-in GPS, this will, generally speaking, work as shown previously, without having to do anything else. If you have an external GPS, you will need to install the drivers and software and write some more code to translate the incoming coordinates into something ArcGIS Runtime can use.

The `AutoPanMode` method includes three options: `Off`, `Default`, and `Navigation`. The `Off` option indicates that the map does not move as the device moves. The `Default` option re-centers the map to the current location. Lastly, `Navigation` changes the heading and position of the map as the location and heading changes for the device. The `CourseSymbol` instance is the current location when moving, while `DefaultSymbol` is the location when stationary. The `IsPingAnimation` property shows an animation as new locations are received.

Another important class is `LocationInfo`, which provides you with a set of events to handle as your location changes. The `LocationChanged` event is used to get new locations as they come from the provider (GPS). With this class, you can get the course, horizontal accuracy, location, and speed. This would be easy to implement using the MVVM pattern with a behavior by handling the `LocationChanged` event, as shown here:

```
AssociatedObject.LocationDisplay.LocationProvider.LocationChanged
    +=
    LocationProvider_LocationChanged;
```

Then, `OnLocationChanged` can be handled:

```
void LocationProvider_LocationChanged(object sender,
    Esri.ArcGISRuntime.Location.LocationInfo e)
{
    this.heading = e.Course.ToString();
    this.speed = e.Speed.ToString();
    this.x = e.Location.X.ToString();
    this.y = e.Location.Y.ToString();
    this.mapView.SetViewAsync(e.Location);
}
```

If you need to use an external GPS, you may need to implement your own LocationProvider class by implementing ILocationProvider. For example, many GPS support the **National Marine Electronics Associated (NMEA)** specification. This specification allows computers to send information between a GPS and a computer. Navigate to http://www.nmea.org/ for more information. Esri has implemented a library on GitHub that allows you to interface with an NMEA compliant device so that you don't have to implement this interface yourself. Navigate to https://github.com/dotmorten/nmeaparser. For your convenience, the DLL has been placed under C:\ArcGISRuntimeBook\NmeaParser.WinDesktop.dll. Once you've added the DLL to your project, you will need to create a class like the following code:

```csharp
public class NmeaLocationProvider : ILocationProvider
{
    public event
        EventHandler<Esri.ArcGISRuntime.Location.LocationInfo>
            LocationChanged;
    private NmeaParser.NmeaDevice device;

    public NmeaLocationProvider(NmeaParser.NmeaDevice device)
    {
        this.device = device;
        device.MessageReceived += device_MessageReceived;
    }

    void device_MessageReceived(object sender,
        NmeaParser.NmeaMessageReceivedEventArgs e)
    {
        var message = e.Message;
        if (message is NmeaParser.Nmea.Gps.Gprmc)
        {
            var rmc = (NmeaParser.Nmea.Gps.Gprmc)message;
            if (rmc.Active && LocationChanged != null)
            {
                LocationChanged(this, new
                    Esri.ArcGISRuntime.Location.LocationInfo()
                    {
                    Course = rmc.Course,
                    Speed = rmc.Speed,
                    Location = new
                    Esri.ArcGISRuntime.Geometry.MapPoint(
                        rmc.Longitude,
                        rmc.Latitude,
                        SpatialReferences.Wgs84)
                });
            }
        }
    }
```

```
        }

        public System.Threading.Tasks.Task StartAsync()
        {
            return this.device.OpenAsync();
        }

        public System.Threading.Tasks.Task StopAsync()
        {
            return this.device.CloseAsync();
        }
    }
```

This class accepts the device (`NmeaDevice`), and then handles incoming location events via `device_MessageReceived`. When the message with the location arrives, it is parsed and the new location is calculated. The `StartAsync` method is called once this class is instantiated. You will then need to instantiate this custom provider using the following code:

```
string comport = "COM3";
int baudRate = 4800;
var port = new System.IO.Ports.SerialPort(commPort, baudRate);
    var device = new NmeaParser.SerialPortDevice(port);
    mapView.LocationDisplay.LocationProvider
    = new NmeaLocationProvider(device);
```

After `NmeaLocationProvider` is instantiated, you just need to set `mapViewLocationDisplay.IsEnabled` = `true` and your external GPS will send in its location and show up on the map.

Summary

In this chapter, we've covered multiple ways to show information on the map so that your users can get a real sense of the data they are seeing and so that the map is interactive. We've gone over ways to handle mouse event using MVVM in such a manner that supports SoC. We've also discussed labeling, adding a legend and scale bar, and ways of working with map tips (overlays) by either using early or late binding. Finally, we've seen ways to show your location, which is a must for any modern mapping system.

In the next chapter, we're going to explain in more detail how to find, query, and identify features on the map using tasks that allow us to execute SQL-like expressions on layers.

7
Finding, Querying, and Identifying Features

When users use your app, they are using it to retrieve information by specifying some criteria or clicking on the map/scene, and then expecting some kind of result that either provides them with the information they asked for, or lets them know that nothing was found. In this chapter, we're going to turn our attention to retrieving information from the layers we've been using so that our users can quickly and easily get information out of the app. In particular, we're going to discuss the options available in ArcGIS Runtime that allow us to find, query, and identify features in a `FeatureLayer` resource. We're also going to discuss the difference between `Find`, `Query`, and `Identify`, how tasks are relevant to this discussion, and how to implement these tasks following the MVVM pattern. We will also discuss the following topics:

- Tasks
- Finding features
- Attributes and spatial queries
- Identify

Tasks

Everything you will be using in this chapter will be related to the .NET concept of tasks. Tasks were introduced in .NET Framework 4 and have greatly simplified the process of writing asynchronous applications, which makes your apps more responsive and non-blocking. A user can enter some search parameter, such as a name, press a button or hit the *Enter* key, and then the operation will do its work in such a manner that it doesn't prevent the user from continuing to work on other parts of the app. ArcGIS Runtime's capabilities are heavily based on this pattern, so it's important that you have a good understanding of how to execute code using this pattern.

In this chapter, we're going to explore `Esri.ArcGISRuntime.Tasks`. This namespace has many classes in it, but for now we're going to just focus on the ones relevant to the topic of this chapter. In later chapters, we'll return to this namespace because the pattern of using tasks is so prevalent when using ArcGIS Runtime. In `Esri.ArcGISRuntime.Tasks`, you'll find the following classes:

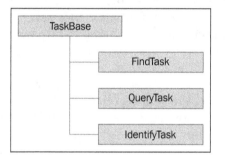

The `TaskBase` class is a base class that has two properties: `ServiceUri` and `Token`. The `FindTask`, `QueryTask` and `IdentifyTask` constructors inherit from `TaskBase`. To use these tasks, we will generally follow this pattern:

1. Instantiate the task using an online or local resource such as `FeatureService` or `FeatureLayer`.

2. Provide the task with the necessary parameters for it to operate. Without some parameters, the task won't return what you need, so this is a necessary step.

3. Now that you have a task and the provided parameters, execute the task.

4. Execute the task using a method such as `ExecuteAsync`, and then do something with the results, such as show the user a list, generate a new geometry, and then display it as a graphic, and send the results to another ViewModel as part of the input to another tool.

Find versus Query versus Identify

The `FindTask` object is used to search one or more layers in a very similar fashion to how a database is searched using an SQL statement. For example, a database is searched using an SQL statement with a WHERE clause, like this:

```
USE MyDatabase;
GO
SELECT Name, ProductNumber, ListPrice AS Price
FROM Production.Product
WHERE ProductLine = 'R'
AND DaysToManufacture < 4
ORDER BY Name ASC;
GO
```

The `FindTask` constructor works with a single WHERE clause against these layers using a text literal. `QueryTask`, on the other hand, can use multiple fields and field types against a single layer. However, it goes beyond that with `QueryTask`. The `QueryTask` class can also be used for spatial queries. For example, you can also execute a query that states "find a parking meter that is on Baker Street and within 500 meters of a grocery store." In effect, `QueryTask` is a tool for doing spatial analysis. As such, this feature makes it a very powerful tool. Lastly, `IdentifyTask` is used to allow the user to interactively click on the map and retrieve information on the layers' features that they clicked on. Think of `Identify` as a drill-down operation. Whatever you click, the fields and their values for all layers will be returned.

Online searching

In this section, we'll go over searching content from online layers.

FindTask

In *Chapter 1, Introduction to ArcGIS Runtime*, we were introduced to a tool that searched through several layers using a `FindTask` constructor. We didn't really discuss how it worked, but we made some changes to it and just enjoyed when it searched through several layers. Let's go over that code in more detail. Here's the code again:

```
var findTask = new FindTask(new System.Uri(this.USAUri));

var findParameters = new FindParameters();
findParameters.LayerIDs.Add(0); // Cities
findParameters.LayerIDs.Add(3); // Counties
```

```
findParameters.LayerIDs.Add(2); // States

findParameters.SearchFields.Add("name");
findParameters.SearchFields.Add("areaname");
findParameters.SearchFields.Add("state_name");

findParameters.ReturnGeometry = true;

SpatialReference sr = new SpatialReference(wkid);
findParameters.SpatialReference = sr;


findParameters.SearchText = this.SearchText;
findParameters.Contains = true;

FindResult findResult = await
    findTask.ExecuteAsync(findParameters);

var foundCities = 0;
var foundCounties = 0;
var foundStates = 0;

// Loop thru results; count the matches found in each layer
foreach (FindItem findItem in findResult.Results)
{
    switch (findItem.LayerID)
    {
        case 0: // Cities
            foundCities++;
            break;
        case 3: // Counties
            foundCounties++;
            break;
        case 2: // States
            foundStates++;
            break;
    }
}

// Report the number of matches for each layer
var msg = string.Format("Found {0} cities, {1} counties, and {2}
    states containing '" + this.SearchText +
    "' in a Name attribute", foundCities, foundCounties,
    foundStates);
```

```
// Bind the results to a DataGrid control on the page
IReadOnlyList<FindItem> temp = findResult.Results;

ObservableCollection<FindItem> obsCollection = new
    ObservableCollection<FindItem>();
foreach (FindItem item in temp)
{
    obsCollection.Add(item);
}

this.GridDataResults = obsCollection;

// show message
Messenger.Default.Send<NotificationMessage>(new
    NotificationMessage(msg));
```

The first thing that happens with this code is the `FindTask` constructor is instantiated with an online service. After that, the most important part of this algorithm is `FindParameters`. The `FindParameters` class allows you to specify which layers to search though using `LayerIDs`. As we discussed earlier, a map service can have one or more layers in it. You specify those layers using their ID, which is always an integer. If you don't know the layers, inspect the service to determine the layers that are in the service and their unique IDs. After the layers are specified, the fields to search against are added to `SearchFields`. You can specify the field names as shown previously, or you can instantiate a generic `List<string>` instance as shown here:

```
System.Collections.Generic.List<string> searchFields = new
    System.Collections.Generic.List<string>();
searchFields.Add("areaname");
findParameters.SearchFields = searchFields;
```

You can also specify the fields using a comma-separated list like so:

```
findParameters.SearchFields.AddRange(new string[] { "CITY_NAME",

    "NAME", "SYSTEM", "STATE_ABBR", "STATE_NAME" });
```

If you don't specify `SearchFields`, all fields will be searched, and that can reduce performance, especially in large layers. After the fields are set, you really need to set the coordinate system so that it matches the coordinate system of `MapView`. The next parameter is the text you want to search for. If you set `Contains` to `true`, an SQL-like expression is used that matches any part of the string literal the user specifies. If `Contains` is set to `false`, it will explicitly look for any record that exactly matches the text that the user enters.

Some other properties worth discussing are `ReturnGeometry` and `LayerDefinitions`. The `ReturnGeometry` property is a Boolean value that indicates whether to return the geometry or not. This option is helpful if you want to do something with the geometry of the results. The `LayerDefinitions` property is a `get` or `set` property that allows you to determine whether the layers had an attribute definition used against them when the layers were originally published. For example, if someone had all of the countries in the world, but only wanted to show European countries, they would have set an expression in `ArcMap` via a definition query (`CONTINENT = 'Europe'`) that only shows those countries. When the layer is published, ArcGIS Server will honor that expression as it is stored with the layer. You can get or change the expression with `LayerDefinitions`, using an example such as this:

```
findParameters.LayerDefinitions = new LayerDefinition[]
{
    new LayerDefinition() { LayerID = 0, Definition =
        "COUNTRY='Norway" },
    new LayerDefinition() { LayerID = 2, Definition =
        "COUNTRY='France" }
};
```

Once the parameters are specified, they are passed into the `FindTask` constructor:

```
FindResult findResult = await
    findTask.ExecuteAsync(findParameters);
```

The `FindTask` constructor will execute the task asynchronously, and then return a `FindResult` class. The `FindResult` class contains the results of `FindTask` and must be awaited. The `FindResult` class has two properties worth further discussion: `ExceededTransferLimit` and `Results`. The `ExceededTransferLimit` property is a Boolean value that simply lets you know whether the number of records returned exceeds the maximum number of records that ArcGIS Server has been set to return when querying a layer. The default value in ArcGIS Server is 1,000 records. Generally speaking, you should check this just to make sure that you haven't exceeded this value, not only for performance reasons, but also because if you exceed this value only 1,000 records will be returned. In essence, if your query returned 1,500 records, but ArcGIS Server is set to 1,000, your query will not have the other 500 records. As a result, it's important that you communicate with your ArcGIS Server administrator to let them know that this setting needs to be higher than the default value.

The `Results` property is the most important object you will be using after `FindTask` finishes, because it has what you're interested in: data. As shown in the preceding code, `ExecuteAsync` returns `Results`, which you can use to get the total number of records, iterate over to present to the user, summarize as shown in the code, and many other useful techniques. The `Results` property contains a collection of `FindItem`, which, as shown previously, you can iterate over using a `foreach` statement. Once the data is summarized, it is then translated into an `ObservableCollection` object so that the `View` can bind to the data.

Canceling a task

The `ExecuteAsync` method also has an overload that includes a `System.Threading.CancellationToken` parameter, which allows you to cancel a task using code such as this:

```
System.Threading.CancellationTokenSource canceller = new
    System.Threading.CancellationTokenSource();
canceller.CancelAfter(10000); // 10 seconds
var token = canceller.Token;

// Execute the task with the cancellation token; await the result
FindResult findResult = null;
Task<FindResult> task = findTask.ExecuteAsync(findParameters,
    token);
try
{
    findResult = await task;
}
catch (System.Threading.Tasks.TaskCanceledException exp)
{
    // ... handle cancel here ...
}
```

This code will simply stop running if the `FindTask` property takes longer than 10 seconds, and then you could let your user know that the query took too long.

QueryTask

As noted earlier, `QueryTask` is more powerful than `FindTask` because it allows you to include spatial queries. But before we delve into spatial queries, let's go over a simple example to show how similar `QueryTask` is to `FindTask`. Before you set up a `QueryTask` property, you must first define a `Query` task. This is similar in concept to `FindParameters`, but with a lot more options.

A basic example of `Query` looks like the following code:

```
Esri.ArcGISRuntime.Tasks.Query.Query query = new
    Esri.ArcGISRuntime.Tasks.Query.Query(sqlQuery);
query.Geometry = this.mapView.Extent;
query.OutSpatialReference = this.mapView.SpatialReference;
query.OutFields.Add("*");
```

The `Query` task has four overloads, which allow you to either pass an `IEnumerable` method of object IDs in a layer or table, a `WHERE` clause, `TimeExtent`, or a `Geometry` class with the spatial relationship. Let's first discuss the object IDs. An object ID is simply a unique ID for each feature in a `FeatureLayer` resource. Every record in a `FeatureService` or `FeatureLayer` resource will automatically have a unique object ID. A `WHERE` clause is the same as discussed earlier with a `FindTask` property. A `TimeExtent` instance is a time window, such as 5 minutes ago:

```
var timeWindow = new Esri.ArcGISRuntime.Data.TimeExtent
(DateTime.Now.Subtract(new TimeSpan(0, 5, 0, 0)), DateTime.Now);
    // 5 minutes ago to present
var queryParams = new
    Esri.ArcGISRuntime.Tasks.Query.Query(timeWindow);
```

The last option we will discuss here is the `Geometry` and spatial relationship overload. With this overload, you can pass in geometry and query the layer using a spatial relationship. For example, you can pass in a `MapPoint` class and use it to see if it intersects with a polygon:

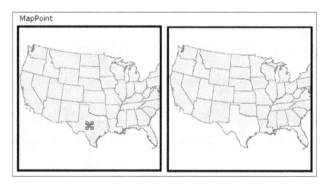

In the preceding screenshot, the point intersects with the state of Texas, so in this case, the state would be returned in the query. The other spatial relationships available include `Contains`, `Crosses`, `EnvelopeIntersects`, `IndexIntersects`, `Overlap`, `Touches`, `Within`, and `Relation`.

Once you've set up the `Query` object, it can be passed into the `QueryTask` constructor, as shown here:

```
Query query = new Query("areaname = '" + this.SearchText + "'");
query.OutFields.Add("*");
System.Uri uri = new System.Uri(this.USAUri + "/0");
QueryTask queryTask = new QueryTask(uri);
QueryResult queryResult = await queryTask.ExecuteAsync(query);
```

Note that this code is using the same layer we used in earlier chapters. In this case, we're querying the cities layer (layer 0) using the `areaname` field. However, with `QueryTask`, you can build more complicated queries such as this:

```
query.Where = "(TYPE = 2 AND STATUS = 1) OR (SPEED <= 10)";
```

Once we execute the `QueryTask` constructor, it returns a `QueryResult` class. The most important thing to note about the difference between `FindTask` and `QueryTask` is that `QueryTask` works on a single layer, not multiple layers. Also, `QueryResult` has a property called `FeatureSet`, which contains a set of features. The `Features` class is a base class to `graphics`, but you can add these features to a `GraphicsLayer` class using code such as this:

```
// Execute the task and await the result
QueryResult queryResult = await queryTask.ExecuteAsync(query);

// Get the list of features (graphics) from the result
var resultFeatures = queryResult.FeatureSet.Features;

// Display result graphics
graphicsLayer.GraphicsSource = resultFeatures;
```

You also have the option of whether or not to return the geometry by setting `ReturnGeometry` to `true` or `false` for `Query`. With `Query`, you can also set the `GeometryPrecision` property, which indicates the precision of the geometry. Also, you can set the `OrderByFields` property, which works just like `ORDER BY` in an SQL database. Another useful property is `OutStatistics`, which returns statistics about the field values. With `OutStatistics` you can compute the average, count, max, min, standard deviation, sum, and variance. You can use `OutStatistics` on a layer using an example such as this:

```
SpatialReference sr = new SpatialReference(wkid);

Query queryCityStats = new Query("pop2000 > 1000");
queryCityStats.OutSpatialReference = sr;
queryCityStats.OutFields.Add("*");
queryCityStats.OutStatistics = new List<OutStatistic> {
```

```
        new OutStatistic(){
            OnStatisticField = "pop2000",
            OutStatisticFieldName = "citysum",
            StatisticType = StatisticType.Sum
        },
        new OutStatistic(){
            OnStatisticField = "pop2000",
            OutStatisticFieldName = "cityavg",
            StatisticType = StatisticType.Average
        }
    };
    QueryTask queryTaskCityStats = new QueryTask(new
        System.Uri(this.USAUri + "/0"));
    QueryResult queryResultCityStatus = await
        queryTaskCityStats.ExecuteAsync(queryCityStats);

    IReadOnlyList<Feature> featuresCityStats =
        queryResultCityStatus.FeatureSet.Features;
    Feature feature = featuresCityStats[0];
    double sum = (double)feature.Attributes["citysum"];
    double avg = (double)feature.Attributes["cityavg"];
```

In this example code, an OutStatistics property is created using a generic List instance of the OutStatistic type on the field called pop2000. The average and sum of the population of all cities over 1,000 are calculated. The OutStatisticFieldName instance is set to citysum and cityavg. Using the results of the QueryTask constructor, the attribute values are cast to a couple of variables of the type double.

A quick example using QueryTask

Let's create an app as we did earlier, where the user was able to search for a city, state, or county using a string literal. However, instead of walking you through all of the steps, you can create a project from scratch or just check out the provided sample project called Chapter7. In those early chapters, we used a FindTask constructor, so let's refactor and use a QueryTask constructor instead. You won't need a DataGrid control. Also, instead of reminding you of everything you need to do to set up this project, we will skip that step and let you build the project based on what you've learned in earlier chapters:

1. Create a UI that allows the user to enter in a name and a button that allows them to search. Use the following code:

    ```
    public async void Search(int wkid)
    {
    ```

```csharp
// Create the symbol
SimpleMarkerSymbol markerSymbol = new
    SimpleMarkerSymbol();
markerSymbol.Size = 25;
markerSymbol.Color = Colors.Red;
markerSymbol.Style = SimpleMarkerStyle.Diamond;

SimpleFillSymbol sfsState = new SimpleFillSymbol()
{
    Color = Colors.Red,
    Style = SimpleFillStyle.Solid
};
SimpleFillSymbol sfsCounty = new SimpleFillSymbol()
{
    Color = Colors.Red,
    Style = SimpleFillStyle.Solid
};

SpatialReference sr = new SpatialReference(wkid);

Query queryCity = new Query("areaname = '" +
    this.SearchText + "'");
queryCity.OutSpatialReference = sr;
queryCity.OutFields.Add("*");
QueryTask queryTaskCity =
    new QueryTask(new System.Uri(this.USAUri + "/0"));
QueryResult queryResultCity = await
    queryTaskCity.ExecuteAsync(queryCity);

Query queryStates =
    new Query("state_name = '" + this.SearchText +
        "'");
queryStates.OutSpatialReference = sr;
queryStates.OutFields.Add("*");
QueryTask queryTaskStates =
    new QueryTask(new System.Uri(this.USAUri + "/2"));
QueryResult queryResultStates = await
    queryTaskStates.ExecuteAsync(queryStates);

Query queryCounties = new Query("name = '" +
    this.SearchText + "'");
queryCounties.OutSpatialReference = sr;
queryCounties.OutFields.Add("*");
QueryTask queryTaskCounties =
    new QueryTask(new System.Uri(this.USAUri + "/3"));
```

```
QueryResult queryResultCounties= await
    queryTaskCounties.ExecuteAsync(queryCounties);

// Get the list of features (graphics) from the result
IReadOnlyList<Feature> featuresCity =
    queryResultCity.FeatureSet.Features;
foreach (Feature featureCity in featuresCity)
{
    Graphic graphicCity = (Graphic)featureCity;
    graphicCity.Symbol = markerSymbol;
    this.graphicsLayerCity.Graphics.Add(graphicCity);
}

// Get the list of features (graphics) from the result
IReadOnlyList<Feature> featuresStates =
    queryResultStates.FeatureSet.Features;
foreach (Feature featureState in featuresStates)
{
    Graphic graphicState = (Graphic)featureState;
    graphicState.Symbol = sfsState;
    this.graphicsLayerState.Graphics.Add(graphicState);
}
// Get the list of features (graphics) from the result
IReadOnlyList<Feature> featuresCounties =
    queryResultCounties.FeatureSet.Features;
foreach (Feature featureCounty in featuresCounties)
{
    Graphic graphicCounty = (Graphic)featureCounty;
    graphicCounty.Symbol = sfsCounty;
    this.graphicsLayerCounty.Graphics.
        Add(graphicCounty);
}
}
```

2. Run your app, and then zoom in around the eastern U.S.

You will see something like the following screenshot if you enter Lancaster as the search string:

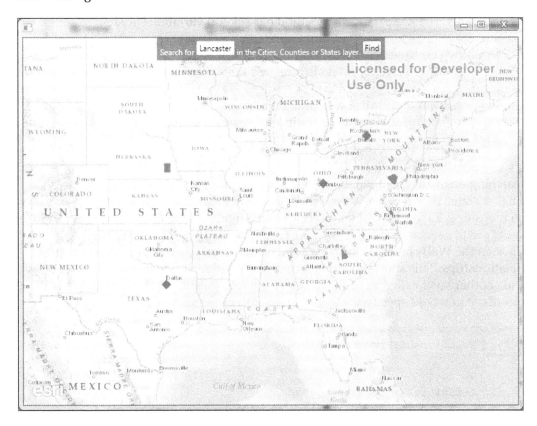

Note that any city, state, or county with the name Lancaster was found and shown on the map. There were no states with the name Lancaster in them, but there were several cities (green triangles) and counties (filled in red) with that name.

The first thing you'll note is that three separate QueryTask objects were created. Then, we retrieved the features, converted them to graphics, and then added them to the GraphicsLayer class. You now know how to create the GraphicsLayer class.

The QueryTask options

The `QueryTask` option is a very powerful search object because it simply provides a lot of options. The `ExecuteAsync` method has several variations (`ExecuteCountAync`, `ExecuteObjectIDsQueryAsync`, and `ExecuteRelationshipQueryAsync`), which mainly allow you to search for the number of features in a layer, search and return just object IDs, and search for data in a related table. For best performance, search using `QueryTask` with `ExecuteObjectIDsQueryAsync`. This method will only return Object IDs instead of the attributes and geometry, so it is inherently faster. With these Object IDs, you can perform additional operations, such as placing the features into a selected state.

It's also possible to search for tables or layers that are related to the layer you're searching on, using `ExecuteRelationshipQueryAsync`. This method expects a `RelationshipParameter` class, which is another class for defining the relationships. Take a look at map service by navigating to `http://sampleserver3.arcgisonline.com/ArcGIS/rest/services/Petroleum/KSPetro/MapServer`. Click on the **Wells** layer. Near the bottom of the page, you'll see a section called **Relationships**. This shows that this layer has a relationship to a table called `Tops` and another layer called `Fields`. To set up the relationship so that you can return records, you would specify the `RelationshipParameter` class similar to this:

```
RelationshipParameter relationshipParameters = new
    RelationshipParameter()
{
    ObjectIds = (int[])ObjIds,
    DefinitionExpression =  exp,
    OutFields = new string[] { "OBJECTID, KID, FORMATION" },
    RelationshipId = 1,
    OutSpatialReference = MyMap.SpatialReference
};
```

In this example, the Object IDs can come from a query or by clicking on a single feature using an `Identify` operation (discussed later in this chapter); then you must specify the `DefinitionExpression` property (the WHERE clause), the output fields, relationship ID, and spatial reference. The relationship ID is in parentheses, as shown here:

> **Relationships:**
>
> • Tops 2 Well (3) -- *Related To:* Wells (0)

In this example, the relationship ID is 3. The `RelationshipParameter` class returns a `RelationshipResult` object, which you can use to iterate over.

Offline searching

In this section, we're going to discuss queries and spatial queries using local content in the Runtime geodatabase.

Querying local layers

Not only can you search online content, but you can also search through an ArcGIS Runtime geodatabase. However, in order to accomplish this, you can't use `QueryTask` or `FindTask` because they require a URI to an online data source. Also, you will need to access the local geodatabase's table instead of directly accessing the layer. To get access to the table, we must first open the ArcGIS Runtime geodatabase. Let's look at an example:

```
var gdb = await Geodatabase.OpenAsync(this.GDB);

Envelope extent = null;
foreach (var table in gdb.FeatureTables)
{
    var flayer = new FeatureLayer()
{
    ID = table.Name,
    DisplayName = "Parking Meters",
    FeatureTable = table
};
```

You've seen this code before. It simply opens a Runtime geodatabase (SQL Lite), and then creates a `FeatureLayer` resource. Let's create some code to search for a parking meter:

1. Create a new ArcGIS Runtime app with MVVM Light and `Json.NET`, or take a look at the sample that came with this book called `Chapter7a`.

2. Add the following XAML code:

```
<TextBlock  Name="Search" Background="#77000000"
    HorizontalAlignment="Center"
    VerticalAlignment="Top" Padding="5"
    Foreground="White" >
    <Run>Enter a parking meter ID:   </Run>
    <TextBox Name="SearchTextBox" Width="50" Text="{Binding
        SearchText}"></TextBox>

    <Button Content="Find" Width="50" Command="{Binding
        SearchRelayCommand}"
        CommandParameter="{Binding Path=Text,
```

```
        ElementName=SearchTextBox}" >
    </Button>
</TextBlock>
```

3. Add the following private member `RelayCommand` to the `MainViewModel` class:

```
public RelayCommand<string> SearchRelayCommand { get;
    private set; }
```

4. In the constructor, instantiate the `RelayCommand` member:

```
this.SearchRelayCommand = new RelayCommand<string>(Search);
```

5. Create a method that takes in the search text and perform the search:

```
public async void Search(string searchString)
{
    FeatureLayer featureLayer = this.mapView.Map.Layers[1]
    as
        FeatureLayer;
    GeodatabaseFeatureTable table =
        featureLayer.FeatureTable as
        GeodatabaseFeatureTable;

    // Define an attribute query
    var filter = new Esri.ArcGISRuntime.Data.QueryFilter();
    filter.WhereClause = "POST_ID = '" +
    searchString + "'"; // 666-13080

    // Execute the query and await results
    IEnumerable<Feature> features = await
        table.QueryAsync(filter);

    foreach(Feature feature in features)
    {
        string address = feature.Attributes["STREETNAME"]
            as string;
        System.Diagnostics.Debug.WriteLine("Address: " +
            address);
    }

}
```

6. Run the app and enter 666-13080 as the meter ID. The address will be sent to the Visual Studio **Output** window in this example.

As you can see from this code, the process is still quite similar to using a `QueryTask` class, except that we had to get the table of `FeatureLayer`. We then set a `QueryFilter` object using a `WHERE` clause. In this particular layer, the meter's ID is called `POST_ID`. It is of type string. Once the query is finished, it returns an enumerable list of features, which we iterated through to get the address in this case. One significant piece of information to be aware of is that querying the ArcGIS Runtime geodatabase requires that you follow the syntax rules of SQLite. For more information, navigate to `http://www.sqlite.org/docs.html`.

Spatial querying local layers

Let's do something a little more interesting with offline data in this section. In this section, we're going to search for all the meters within 200 meters of meter `666-13080`. To implement this, add a new method to your `ViewModel` class called `SearchByMeterID`. It will have the same signature as the previous searching method:

1. Add the following method:

```
public async void SearchByMeterID(string searchString)
{
    SimpleLineSymbol sls = new SimpleLineSymbol()
    {
        Color = System.Windows.Media.Colors.Red,
        Style = SimpleLineStyle.Solid,
        Width = 2
    };

    // get the layer and table
    FeatureLayer featureLayer = this.mapView.Map.Layers[1]
        as
        FeatureLayer;
    GeodatabaseFeatureTable table =
        featureLayer.FeatureTable as
        GeodatabaseFeatureTable;

    // Define an attribute query
    var filter = new Esri.ArcGISRuntime.Data.QueryFilter();
    filter.WhereClause = "POST_ID = '"
        + searchString + "'"; // Try 666-13080

    // Execute the query and await results
    IEnumerable<Feature> features = await
        table.QueryAsync(filter);
```

```
// iterate the feature. Should be one in this case.
foreach (Feature feature in features)
{
    // Get the MapPoint, Project to Mercator so that we
        are
    // working in meters
    MapPoint mapPoint = feature.Geometry as MapPoint;
    MapPoint pointMercator =
        GeometryEngine.Project(mapPoint,
        SpatialReferences.WebMercator) as MapPoint;
    Geometry polygon =
        GeometryEngine.Buffer(pointMercator, 200);

    // Re-project the polygon to WGS84 so that we can
        query
    // against the layer which is in WGS84
    Polygon polygonWgs84 =
        GeometryEngine.Project(polygon,
        SpatialReferences.Wgs84) as Polygon;

    // add the circle (buffer)
    Graphic graphic = new Graphic();
    graphic.Symbol = sls;
    graphic.Geometry = polygonWgs84;
    this.graphicsLayer.Graphics.Add(graphic);

    // Make sure the table supports querying
    if (table.SupportsQuery)
    {
        // setup the query to use the polygon that's in
            WGS84
        var query = new SpatialQueryFilter();
        query.Geometry = polygonWgs84 as Geometry;
        query.SpatialRelationship =
            SpatialRelationship.Intersects;

        var result = await table.QueryAsync(query);

        // Loop through query results
        foreach (Esri.ArcGISRuntime.Data.Feature f in
            result)
        {
            // do something with results
        }
```

```
                }
            }
        }
```

2. Update the `RelayCommand` and XAML code to find this new method.

3. Run the app and enter a parking meter ID of `666-13080`.

You will see a buffer on the map around the meter found:

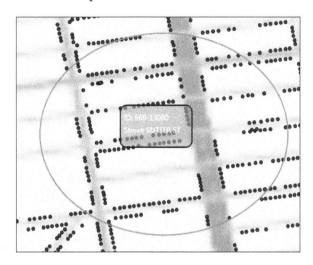

In this method, we've set up a `SimpleLineSymbol` class so we can view the buffer, retrieved the layer and layer's table, defined a `QueryFilter` object so that we can get the `MapPoint` instance of a parking meter, and iterated over the parking meters. Once we have a parking meter, we re-project the parking meter's `MapPoint` class from WGS84 to Mercator so that we can buffer it using a hardcoded value of 200 meters. We can't use meters with a WGS84 coordinate system because that coordinate system uses angular units. We then compute the buffer, which produces a polygon, and then we re-project the polygon back into WGS84 so that it will display correctly on the map. After that, we create a `Graphic` class of the buffer and display it on the map. Then, we check to make sure that the table supports being queried. This is an optional setting when the layer was published to a ArcGIS Runtime geodatabase. Then, we set up the `SpatialQueryFilter` class to use the polygon in WGS84, and set up the spatial relationship to search for any parking meter that intersects with the polygon. Lastly, we issue `Query` asynchronously, and then iterate over the results.

This example shows how to perform queries on offline data, while at the same time it shows how to perform spatial analysis. Spatial analysis is the primary engine of GIS and we will return to it in a later chapter.

IdentifyTask

Imagine for a moment that you have multiple layers in your map or scene, and you want to click on a particular location and find everything at that location. See the following image:

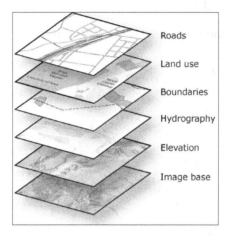

If you click on these layers, the IdentifyTask object will return every feature in every layer that is located where you clicked. The IdentifyTask class uses the concept of a hit test. A hit test simply means that if the point where the user clicks intersects with any feature in any layer, those features are returned to IdentifyTask. It's possible to perform an Identify operation on both online and offline data. Let's look at an online example first, and then explore this further with the same kind of operation but with offline data.

All examples in this section are in the code provided with this book, which is called Chapter7a.

The online Identify task

Let's add some code, and then discuss how IdentifyTask works using an online service:

1. In your project, open the MapViewBehavior.cs file and add the following using statement: using Esri.ArcGISRuntime.Tasks.Query;. Then, add a new event to the OnAttached method, as shown here:

    ```
    AssociatedObject.MouseUp += AssociatedObject_MouseUp;
    ```

2. In the event handler, add the following code:

```
async void AssociatedObject_MouseUp(object sender,
    System.Windows.Input.MouseButtonEventArgs e)
{
    MapView mapView = sender as MapView;
    var screenPoint = e.GetPosition(mapView);

    // Convert the screen point to a point in map
        coordinates
    var mapPoint = mapView.ScreenToLocation(screenPoint);

    // Create a new IdentifyTask pointing to the map
        service to
    // identify (USA)
    var uri = new
        Uri("http://sampleserver6.arcgisonline.com/arcgis/
            rest/services/USA/MapServer");
    var identifyTask = new IdentifyTask(uri);

    // Create variables to store identify parameter
        information
    //--current map extent (Envelope)
    var extent = mapView.Extent;
    //--tolerance, in pixels, for finding features
    var tolerance = 7;
    //--current height, in pixels, of the map control
    var height = (int)mapView.ActualHeight;
    //--current width, in pixels, of the map control
    var width = (int)mapView.ActualWidth;

    // Create a new IdentifyParameter; pass the variables
        above to
    // the constructor
    var identifyParams = new IdentifyParameters(mapPoint,
        extent,
    tolerance, height, width);

    // Identify only the top most visible layer in the
        service
    identifyParams.LayerOption = LayerOption.Top;

    // Set the spatial reference to match with the map's
    identifyParams.SpatialReference =
        mapView.SpatialReference;
```

```
    // Execute the task and await the result
    IdentifyResult idResult = await
        identifyTask.ExecuteAsync(identifyParams);

    // See if a result was returned
    if (idResult != null && idResult.Results.Count > 0)
    {
        // Get the feature for the first result
        var topLayerFeature = idResult.Results[0].Feature
            as Graphic;

        // do something
    }
}
```

Note that this example first takes where the user clicked and converts it to a screen point. Next, that screen location is converted to a MapPoint class. Then, an online service URI is created and passed into the IdentifyTask constructor. Next, the extent, tolerance (in pixels), extent height, and width are created, and then passed into the IdentifyParameters argument, where several parameters are set up. From there, IdentifyTask is executed and IdentifyResult returned so that something can be done with it.

The IdentifyParameters argument has several options, so let's discuss them in more detail. Other than the MapPoint class, you must specify an extent so that it limits the search area to perform the Identify operation. Next is the tolerance. This is the distance in screen pixels from the specified geometry within which to execute the identify operation. The tolerance is very important because it is dependent on the size of the symbol being used for the features. If the symbol is really small, a larger tolerance needs to be used. If the symbol is very large, a smaller tolerance can be used. Lastly, the screen width and height are also specified.

The LayerOption instance allows you to specify that only the Top, Visible, or All layers will be hit tested for the Identify operation. The All option should be used carefully with map services that have many layers, because it can reduce performance. Also, with the Visible option, only layers that have their visibility set to true will be used in the Identify operation. Lastly, if layers are grouped, you must specify the All option; otherwise, nothing will be returned in the layers in GroupLayer.

The LayerIDs property allows you to specify specific layers to perform the Identify operation on. Think of this property as an AND operation. If layer IDs 0 and 1 are specified and are visible, they will be identified.

The offline Identify task using MVVM Light's messenger

Let's add some code, and then discuss how an identify operation works with offline data:

1. In your project, open the `MapViewBehavior.cs` file and add the following using statements for `Esri.ArcGISRuntime.Tasks.Query` and `Esri.ArcGISRuntime.Data`. Then, add a new event in the `OnAttached` method, as shown here:

   ```
   AssociatedObject.MouseUp += AssociatedObject_MouseUp;
   ```

2. In the event handler, add the following code:

   ```
   MapView mapView = sender as MapView;
   var screenPoint = e.GetPosition(mapView);

   // Convert the screen point to a point in map coordinates
   var mapPoint = mapView.ScreenToLocation(screenPoint);

   // get the FeatureLayer
   FeatureLayer featureLayer = mapView.Map.Layers[1] as
       FeatureLayer;
   // Get the FeatureTable from the FeatureLayer.
   Esri.ArcGISRuntime.Data.FeatureTable featureTable =
       featureLayer.FeatureTable;

   // Translate the MapPoint into Microsoft Point object.
   System.Windows.Point windowsPoint =
       mapView.LocationToScreen(mapPoint);
   // get the Row IDs of the features that are hit
   long[] featureLayerRowIDs =
       await featureLayer.HitTestAsync(mapView, windowsPoint);

   if (featureLayerRowIDs.Length == 1)
   {

       // Cause the features in the FeatureLayer to highlight
           (cyan) in
       // the Map.
       featureLayer.SelectFeatures(featureLayerRowIDs);

       // Perform a Query on the FeatureLayer.FeatureTable
           using the
   ```

```
        // ObjectID of the feature tapped/clicked on in the map.

        // Query the layer using the Row IDs
        IEnumerable<GeodatabaseFeature> geoDatabaseFeature =
            (IEnumerable<GeodatabaseFeature>)await
        featureTable.QueryAsync(featureLayerRowIDs);

        foreach (Esri.ArcGISRuntime.Data.GeodatabaseFeature
            oneGeoDatabaseFeature in geoDatabaseFeature)
        {
            // Get the desired Field attribute values from the
            // GeodatabaseFeature.
            System.Collections.Generic.IDictionary<string,
                object>
                attributes = oneGeoDatabaseFeature.Attributes;

            object postID = attributes["POST_ID"];

            // Construct a StringBuilder to hold the text from
                the Field
            // attributes.
            System.Text.StringBuilder stringBuilder = new
                System.Text.StringBuilder();
            stringBuilder.AppendLine("POST ID: " +
                postID.ToString());

            // Send to messenger
            Messenger.Default.Send<NotificationMessage>(new
            NotificationMessage(stringBuilder.ToString()));
        }
    }
```

This code segment essentially produces the same result as the `IdentifyTask` constructors we used in the previous section, but with one big difference: we didn't use `IdentifyTask`. The reason for this is that `IdentifyTask` requires a URI to an online service, but we're using a local ArcGIS Runtime geodatabase. In fact, in the code we're just getting the second layer in the layers that are in the map. Once we get the `FeatureTable` class of `FeatureLayer`, we perform a hit test using `HitTestAsync` on `MapView` using the Windows point, which is in pixels. This returns a set of row IDs. In this case, we only wanted to perform the hit test on one layer, but we could have easily performed the hit test on whatever layers we wanted (`Top`, `Visible`, or `All`).

Once we have the row IDs, we put them into a selected state. If you zoom in, you'll note that the color of the parking meter has changed to cyan. This means it's selected. This isn't necessary, but it lets the user know what they're identifying. If they missed the right feature, they will immediately know it. Once the feature is selected, we perform a QueryAsync variation using the row IDs. We then iterate over the returned GeodatabaseFeature class and create a dictionary of attributes so that we can retrieve the attribute fields and values. Most importantly, we then sent the attributes in StringBuilder to the MVVM Light messenger we set up in *Chapter 2, The MVVM Pattern*. As a result of using a custom behavior and the messenger, we've satisfied the notion of SoC.

The Identify options

Some other things to keep in mind include how you want to handle overlays and the Identify operations, as shown in this chapter. For example, if you want to perform an Identify operation and overlay, you'll need to figure out which mouse events to use in Windows. Do you use right-click, left-click, or mouse up? This can get even more complicated in Windows Store because you typically will want to handle MapViewTapped. How will you handle an Identify operation and overlay? You could give the user a dialog that provides an option, or you could pick and choose which layers to use for the overlay and which to use for the Identify operation. Also, when designing an Identify tool, you need to think about which fields to show. If your app has a pre-set list of layers with known fields, this would simply be a matter of hardcoding them in, but that's not the most elegant solution. With Runtime, you can read the layer fields and write your code in such a manner to generate the result dynamically. No matter what, careful consideration should be given to how you implement these interactive tools.

Summary

In this chapter, we've learned about tasks, the difference between FindTask, QueryTask, and IdentifyTask, online searching with FindTask and QueryTask, canceling a task, offline searching with local layers, spatial queries, and performing an Identify operation with online and offline data.

In the next chapter, we'll delve into a different kind of searching related to addresses. Once we've learned about searching for a location using an address, we'll turn our attention to routing between addresses.

8

Geocoding and Routing

Now that we've learned how to search for data in layers and related tables, we need to turn our attention to searching by using an address or the name of a place. With ArcGIS Runtime, it's possible to search using an address, such as 123 Main Street, via a process called geocoding. Geocoding is a common feature in modern mapping technologies, so it's important to master the concepts and techniques involved. In this chapter, we'll cover the process of geocoding, and then we'll learn about routing, which determines the shortest route between two locations (address, place, or coordinates). In this chapter, we'll cover the following topics:

- Online geocoding
- Offline geocoding
- Finding places
- Online routing
- Offline routing

Geocoding

Geocoding is the process of taking an address and translating it into coordinates. 10 Main Street, San Francisco, CA 94105 is actually at latitude/longitude 37.792687, -122.396289. An address is composed of the following parts:

Components	Description
Left house number range	This range is includes a low number and high number for the left side of the street, such as 100 to 198.
Right house number range	This range includes a low number and high number for the right side of the street.
Prefix direction	This component is a directory indicator, such as E, W, N, S. For example, W. Maple St.

Components	Description
Prefix type	This component is the type of street that precedes the name, such as Avenue C.
Street name	This component is the name of the street, such as Main or Peach.
Street type	This component is the type of street that follows the name.
Suffix direction	This component is a direction that follows the street name, such as NE in Baker St. NE.
City name	This component is a city name, such as New York.
State	This component is the state name or abbreviation, such as NC.
zip code	This component is the postal code used by the United States Postal Service, such as 84123.

Geocoding in ArcGIS Runtime will take the address and determine the coordinate of that address using an algorithm and underlying street data. Streets, of course, are linear features that have a set of attributes such as street range, street name, city, and state. The important part of this is the street range. A street will typically have a range for the right side and left side:

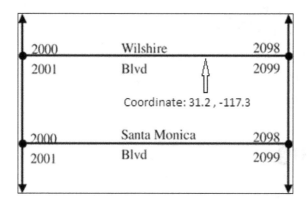

Each street will have the range, street name, city, state, and a ZIP code (postal code). Finding a street using just the street name, city, state, and ZIP code is really no different than what you learned in the previous chapter. It's just an attribute query. The algorithmic part of this process is to determine where the address number sits in the address range. If the range is 100 to 200 and the address is 150, the algorithm will place the point at the middle point of the line.

Locators

It's important to understand that the data source is the key component that is necessary to make geocoding possible. However, to make the algorithm as efficient as possible, an abstraction of the data is created. It's called a locator. The locator's role is to allow you to set the address style used in the underlying street network. A locator style lets the locator know what kind of addressing scheme to use. Here are some example locator styles: dual ranges on both sides of the street, a single range, single house, zip 5-digit, and Gazetter. The reason for this is that the geocoding engine needs to know what kind of addresses you intend to use. The locator style is specified when you create the locator using ArcGIS Desktop or Pro. For more information, navigate to `https://desktop.arcgis.com/en/arcmap/10.3/guide-books/geocoding/understanding-address-locator-styles.htm`. These styles determine the geocoding properties and parameters that determine how the address is parsed, and contain a snapshot of the address attributes in the street data, and the criteria for performing the geocoding operation. Once a user passes in an address, the geocoding engine parses the address into its constituent parts, determines candidate addresses, and then determines the coordinates. The process for creating a locator is explained in the ArcGIS help section located at `https://desktop.arcgis.com/en/desktop/latest/guide-books/geocoding/creating-an-address-locator.htm`.

The locator can be published as a service either via ArcGIS Online, on-premises in ArcGIS Server, or placed locally on a device as a `*.loc` file. Esri has published several locators for the U.S. and Europe. You can also purchase street data from a wide variety of street data vendors and build your own locator. This is a vital option for organizations that need to protect the privacy of their customers, such as for patients in the healthcare industry.

Candidates and scores

Once you've built the locator, you're now ready to geocode, but let's first explain a little more about how the process works. The first thing to do is to pass in an address. The more information you provide, the higher the probability the address will be found. In many cases, you will pass in an address and the geocoding engine will let you know that it has found the **candidate** addresses. It just depends on the preciseness of the entered address and the quality of the underlying street data. For example, if a street has a name of Smith Street with a range of 100 to 8,000, but you entered `8020`, the geocoding engine can only provide a candidate address such as 7999 Smith Street or 8000 Smith Street. You will then need to present these candidate addresses to the user so that they can choose which one they really want. In effect, you're also letting the user know that the address is an error either by the user or in the underlying street data.

It's also important to know that each candidate address will receive a score from 0 to 100. The higher the score, the more likely that it is the correct one to use. In your code, you can sort the address candidates by their score. The locator can also be set up to reject candidates below a certain score. By default, ArcGIS Desktop rejects candidate addresses with a score below 80, so you are guaranteed to have good candidate addresses for the user to review. You can apply the same logic to your app.

The address-matching process actually provides you with even more information about the address you've entered when you attempt to geocode an address. The information returned about the candidate address includes the following values:

Values	Description
Score	This value gives a value between 0 and 100, which indicates the accuracy of the match.
Match_addr	This value gives the complete address returned by the geocoding engine.
Addr_type	This value gives the match level for a geocoding request. For example, if the passed-in address only matches the city, a value called Locality will be placed in this field. For more information, see Addr_type at http://resources.arcgis. com/EN/HELP/ARCGIS-REST-API/INDEX.HTML#/SERVICE_ OUTPUT/02R300000017000000/ESRI_SECTION1_42D7D3D02 31241E9B656C01438209440/.
Side	This value gives the side of the street on which the address was matched. This depends on the direction the street was digitized. There are three possible values: L (left), R (right), or an empty string.
Distance	This value gives the distance in meters from the candidate to the location along the street.
Rank	If two or more address candidates receive the same Score value, the Rank value can help determine the better answer. For example, if the address contains Washington, the geometry engine will choose Washington, DC because it has a greater population than Washington, Nebraska.

With this information, you can write your code to decide whether or not to present the candidate address to the user and how to show the address on the map.

Useful terms

While learning about geocoding, it's important to understand some useful terms:

- **Batch geocoding**: This term means geocoding more than one address at a time
- **Reverse geocoding**: This term means converting a coordinate, such as latitude/longitude, into an address
- **Address matching**: This term is a synonym for address geocoding

In the sections that follow, we will use these terms interchangeably where appropriate.

Locator tasks

As we discussed in the previous chapter, tasks are a very important concept to understand and utilize in order to work with ArcGIS Runtime. Fortunately, what you learned in that chapter applies to this chapter too. To perform geocoding, we will explore the classes located in `Esri.ArcGISRuntime.Tasks.Geocoding`. To geocode an address, you will need to use a `LocatorTask` class. The `LocatorTask` class is an abstract class for `OnlineLocatorTask` and `LocalLocatorTask`, as shown in the following diagram:

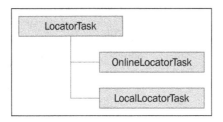

The general process for geocoding one or more addresses is as follows:

1. Create a locator task (`OnlineLocatorTask` or `LocalLocatorTask`) using either an online service or a locator file on the device.

2. If you are looking for a place such as the Coliseum in Rome, instead of using an address, you'll just pass in the name using `OnlineLocatorFindParameters`. Using place names will be discussed shortly.

3. Execute the task asynchronously using the input parameters of an address or `OnlineLocatorFindParameters`.

4. Handle the results. This could be in the form of just showing the location or presenting the user with a list of candidate addresses to choose from.

Online geocoding

To geocode an address using an online service, we will simply use the following code:

```
// specify the geocoding service
var uri = new
    Uri("http://geocode.arcgis.com/arcgis/rest/services/World/
        GeocodeServer");
var locatorTask = new
    Esri.ArcGISRuntime.Tasks.Geocoding.OnlineLocatorTask(uri);
var info = await locatorTask.GetInfoAsync();

// create the name of the single address input field
var singleAddressFieldName =
    info.SingleLineAddressField.FieldName;

// create a dictionary to contain the address info
var address = new Dictionary<string, string>();
// add address components to the dictionary using the correct input
field names (a single address line in this example)
address.Add(singleAddressFieldName, "10 Main Street, San
    Francisco, CA 94105");

// create a list of output fields to include with the candidates
var candidateFields = new List<string> { "Score", "Addr_type",
    "Match_addr", "Side" };

// start the geocode task: pass in the address, out candidate
    fields, out spatial reference, and an empty cancellation token
var task = locatorTask.GeocodeAsync(address, candidateFields,
    this.mapView.SpatialReference, new
    System.Threading.CancellationToken());

IList<LocatorGeocodeResult> results = null;
try
{
    // await the results of the task
```

```
        results = await task;

        // if at least one candidate was found, get its location
            (point)
        if (results.Count > 0)
        {
            var firstMatch = results[0];
            var matchLocation = firstMatch.Location as MapPoint;
            // ... code here to display point on the map ...
        }
    }
    catch (Exception ex)
    {
        var msg = "Exception from geocode: " + ex.Message;

    }
```

The first thing to note about this code is the name of the geocoding service. It's called `World` and we know it's a geocoding service, because at the end of the URL is the word `GeocodeServer`. The next item of interest is `GetInfoAsync`, which returns information about the geocoding service, such as the address fields, candidate fields, and spatial reference of the service. All of these properties provide us with details that allow us to work with the service. `GetInfoAsync` works with online and offline locators.

As we want to geocode a single address, we need to get the `SingleLineAddressField` parameter so that we can pass in a single address. After that, we create a dictionary that will allow us to add the `SingleLineAddressField` parameter with a single address such as 10 Main Street, San Francisco, CA 94105. After this step, we also need to create a list of output field names. We will use those names to inspect the returned address matches. The next step involves calling `GeocodeAsync` with the address, candidate fields, MapView's spatial references, and a cancellation token. Once this task completes, it will return a result of type `LocatorGeocodeResult`. We then try to get the results. If a result is returned, we get the first matching address. Lastly, we get the location and cast it to a `MapPoint` class so that we can then add it to the map.

If you look at this code in the Visual Studio debugger, you'll note that it returns the first result with the highest score. For this particular address, the score was 100. As mentioned earlier, ArcGIS Desktop uses a score of 80 (default) to decide whether to accept or reject the matching address. You can get the score yourself, and then add a little more logic to render the point on the map or not, based on the score.

You can find other online locators at `http://sampleserver1.arcgisonline.com/ArcGIS/rest/services/Locators`.

The `LocatorTask` class can also reverse-geocode a `MapPoint` class using `ReverseGeocodeAsync`. For example, you could have the user click on a street and have it report back with the address.

If you wanted to geocode a batch of addresses, it would be easy enough to add more items to the dictionary in the preceding code and iterate over each one of them. This could be accomplished by allowing the user to upload a spreadsheet, or a CSV file, or whatever. Another alternative is to allow the user to drag and drop a CSV file onto the map, just like with ArcGIS Online.

Offline geocoding

Offline geocoding works very similarly to online geocoding except you must include the locator file with your app and place it on the device. Then, instead of using `OnlineLocatorTask`, you'll use `LocalLocatorTask`, like this:

```
var locatorTask = new
    LocalLocatorTask(@"C:\ArcGISRuntimeBook\Data\
        SanFranciscoLocator\SanFranciscoLocator.loc");
var info = await locatorTask.GetInfoAsync();
```

This locator is included with the sample data provided in this book.

Finding a place

The `OnlineLocatorTask` class provides one capability that `LocalLocatorTask` doesn't provide. You can use `OnlineLocatorTask` to search for places. For example, you can just enter a name, such as `Eiffel` for the Eiffel Tower in Paris, or `Mount Everest`, and the service will take you to this place. The `OnlineLocatorTask` class provides a method called `FindAsync` that allows you to use the following information:

- Street addresses
- Points of interest by name and category
- Administrative place names, such as cities, counties, states, provinces, and country names
- Postal codes

To use the `FindAsync` method, create a `LocatorFindParameters` class. This class allows you to specify which fields to return and the spatial reference, limit the number of candidates with `MaxLocations`, restrict the search area to a particular country with `SourceCountry`, limit the geographic extent with `SearchExtent`, and so on.

Let's look at an example:

```
// create a locator to search the ArcGIS Online World geocode
    service
var uri = new
    Uri("http://geocode.arcgis.com/arcgis/rest/services/World/
        GeocodeServer");
var token = String.Empty;
var locator = new
    Esri.ArcGISRuntime.Tasks.Geocoding.OnlineLocatorTask(uri,
        token);

// get the locator's spatial reference
var info = await locator.GetInfoAsync();
var locatorSpatialRef = info.SpatialReference;

// create find parameters; search locations for "coffee"
var findParams = new OnlineLocatorFindParameters("coffee");

// set the output spatial reference to match the map's
findParams.OutSpatialReference = this.mapView.SpatialReference;

// restrict the search to the current map extent
var searchExtent =.GeometryEngine.Project(this.mapView.Extent,
    locatorSpatialRef);
findParams.SearchExtent = (Envelope)searchExtent;

// prioritize matches within 7.5 kilometers of the center of the
    current map extent
var centerOfMap = this.mapView.Extent.GetCenter();
var centerOfMapLatLong =
    (MapPoint)GeometryEngine.Project(centerOfMap,
        locatorSpatialRef);
findParams.Location = centerOfMapLatLong; // distance to
    candidates will be measured from here
findParams.Distance = 7500; // meters

// return a maximum of 7 candidates
findParams.MaxLocations = 7;

// return phone number, url, and distance attributes with the
    candidates
var candidateFields = new List<string> { "Phone", "URL",
    "distance" };
```

```
findParams.OutFields = candidateFields;

try
{
    // execute a find on the locator using the parameters defined
        above
    var task = locator.FindAsync(findParams, new
        CancellationToken());
    IList<LocatorFindResult>
        results = await task;
    foreach (var candidate in results)
    {
        // get information about each candidate
        var candidateFeature = candidate.Feature;
        var name = candidate.Name;
        var phone =
            candidateFeature.Attributes["Phone"].ToString();
        var url = candidateFeature.Attributes["URL"].ToString();

        // get distance in miles from the center of the map
        var distMeters = 0.0;
        double.TryParse(candidateFeature.Attributes["distance"].
            ToString(), out distMeters);
        var distMiles = distMeters * 0.00062137; // convert from
            meters to miles

        // get location (geometry)
        var matchLocation = candidateFeature.Geometry as MapPoint;

        // ... code here to display point graphics and info ...
    }
}
catch (Exception ex)
{
    var msg = "Exception from geocode: " + ex.Message;
}
.
```

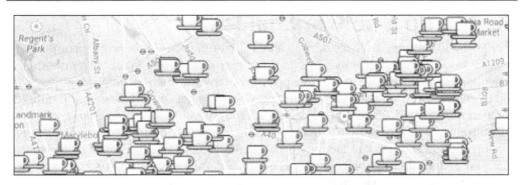

Let's review this code. First, we open the `OnlineLocatorTask` class, but this time we pass in an empty token. Then, the `OnlineLocatorTask` class is instantiated as we've seen before, but this time we create `OnlineLocatorFindParameters` and pass in the word "coffee". We then provide the spatial reference for `OnlineLocatorFindParameters`. After that, we project the extent to the locator's spatial reference so that we can search for coffee shops within a radius of 7.5 km from the center of the map. We also tell `OnlineLocatorFindParameters` to limit its results to just seven candidates. We then specify the phone number, URL, and distance from the map's center. Next, we specify the output fields with `OutFields`. Then, we try to execute `FindAsync`. For each result returned, we get the candidates and return the field values we specified earlier. If the distance is within the search tolerance of 7.5 km, we can do something with the results, such as draw them on the map. We could even create an overlay dynamically and populate it with the URL so that users could click on the URL and take them to the coffee shop's website. The coffee symbols in the preceding screenshot were created using `PictureMarkerSymbol`.

Routing

Although geocoding can be used to map discrete locations with an address, such as coffee shops, it is often the case that geocoding leads to routing. Once you choose the coffee shop that you want to go to, it's then desirable to compute the shortest path so that you can minimize your travel time. Routing is the process of minimizing the distance traveled on a street network. The good news is that routing uses the concept of a task just like the previous tasks we've explored. To compute a route, we're going to use the `RouteTask` class. The `RouteTask` class will find the shortest or fastest route between two or more locations, called **stops**. It goes beyond just traversing the route though. The `RouteTask` class will also take into account restricted areas, and you can even specify a time window that incorporates the arrival and departure times.

Setting up a network dataset

In order to do routing, it is necessary to either use online services or create your own routing service. No matter which option is required, a network dataset is required. A network dataset is a model of the transportation network, such as streets, highways, and freeways. They are derived from the polyline geometry in a dataset that is either purchased or created. Once this street data is brought into ArcGIS Desktop or Pro, it is then translated into a network dataset, which, in a highly optimized manner, stores relationships such as the connectivity between roads. For example, based on the source data's fields and the way the geometry is organized, the network datasets will know that bridges pass over roads and are therefore not connected. See here:

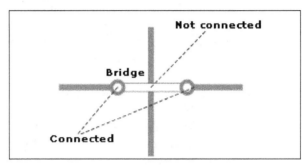

Network datasets

The preceding image is also available at `https://desktop.arcgis.com/en/desktop/latest/guide-books/extensions/network-analyst/GUID-3F65156A-E4F4-4FEA-9A1B-D3AC816746D9-web.gif`.

It will also store information such as the number of turns at an intersection, the speed limit along each segment, the time it takes to traverse a road segment, the hierarchy of the streets (interstates, highways, streets, and so on), and other information that allows for a complete model of the street system. Once the network dataset is built, it can then be published as a service for online use or converted to a ArcGIS Runtime geodatabase for offline use. This service is called a **Network Analyst** service and it can be accessed just like the many other services available from ArcGIS Online. To learn more about network datasets, navigate to `http://resources.arcgis.com/en/help/main/10.2/index.html#//004700000007000000`. To learn how to create a Network Analyst service, navigate to `http://resources.arcgis.com/en/help/main/10.2/index.html#Route_analysis/004700000045000000/`.

An example of an online service is the Network Analyst service available at the following site:

`http://sampleserver6.arcgisonline.com/arcgis/rest/services/NetworkAnalysis/SanDiego/NAServer`

If you open up the service in your favorite browser, you will note that it has **Service Description**. In **Service Description**, it states that this is for the area of San Diego, CA and that it has the capabilities shown in the following screenshot:

If you click on each of these links, you will discover what each kind of analysis service can do. For example, **Route Layers** exposes a routing service (**Route**) which allows you to calculate the route between two or more locations. It also lists a set of parameters, such as **Impedance**, **Restrictions**, **Output Spatial Reference**, and **Network Analysis Classes**. In the sections that follow, we'll explain what these types of analysis mean and how to use them.

Routing overview

In this section, we're going to be using objects in `Esri.ArcGISRuntime.Tasks.NetworkAnalyst`. Let's take a look at the object model diagram:

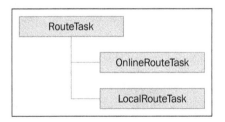

As shown in the diagram, routing tasks work just like the other tasks we've seen. There is an abstract class called `RouteTask`, and under it are `OnlineRouteTask` and `LocalRouteTask`. The only real difference between the two is that `OnlineRouteTask` requires an input URI and `LocalRouteTask` requires a Runtime geodatabase. Here are the two examples of creating a routing task:

```
// Online Routing Task.
string onlineRoutingService =
    "http://sampleserver6.arcgisonline.com/arcgis/rest/services/
        NetworkAnalysis/SanDiego/NAServer/Route";
OnlineRouteTask routeTask = new OnlineRouteTask(new
```

```
        Uri(onlineRoutingService));

// Offline routing task.
LocalRouteTask task = new LocalRouteTask
(@"C:\ArcGISRuntimeBook\Data\Networks\
    RuntimeSanFrancisco.geodatabase", "Streets_ND");
```

Setting up the input parameters

Now that we have a task, we need to pass in the information it requires in order for it to produce a route. The `RouteParameters` class comes to the rescue. The `RouteParameters` class requires a minimum of two stops (the `MapPoint` class) to calculate a route. This is accomplished with `SetStops`. However, as we saw when we looked at the service earlier, it comes with a set of parameters already built in. As such, you can use `GetDefaultParametersAsync` for the default parameters. This can be a time saver if your routing service has been set up to meet your particular needs. Otherwise, you'll need to set the parameters for your particular requirements.

Some of the options you have to consider are impedance, barriers, driving directions, U-turn policy, geometry options, restrictions, and so on. Let's look at some of these options, and then we'll build an app that shows how offline routing is set up and executed using the San Francisco data we've been using in this book.

Stops

The `SetStops` method allows you to specify the stops the routing task will use to create the routes. The `SetStops` method takes two or more `MapPoint` classes in a collection in order to create a route. The order in which the route is generated is based on the order in which you've added the `MapPoint` classes to the collection. However, if you set `FindBestSequence` to `true` on `RouteParameters`, the order will be re-ordered. When `FindBestSquence` is `true`, the routing algorithm will switch from the shortest-path problem to the **traveling salesperson problem** (TSP). We won't cover TSP in this book.

Impedance

Impedance is the cost that needs to be minimized when traversing the streets. Impedance could be distance, time, or some other value such as cost. If you go back and look at the San Diego routing service again, you'll note that the impedance is `TravelTime`, which is a field in the underlying street network. You could change this to another field, such as `Length`. To set this field, use `ImpedanceAttributeName`. If you open the service, you will note, under **Network Attributes**, one or more fields. For any field that is of **Usage Type: esriNAUTCost**, you can use this field name for the impedance. See here:

- **Minutes**
        Usage Type: esriNAUTCost
        Units: esriNAUMinutes
        Data Type: esriNADTDouble
        Restriction Usage ParameterName: null
        Parameter Names:
- **Meters**
        Usage Type: esriNAUTCost
        Units: esriNAUMeters
        Data Type: esriNADTDouble
        Restriction Usage ParameterName: null
        Parameter Names:

The fields called `Minutes` and `Meters` could be used as impedance fields.

Barriers

A barrier is a point, polyline, or polygon that forces the route to navigate around it. In the following screenshot, note that there is an **X** mark on the road between stops **1** and **2**. Because of this point barrier, the route was re-routed around it:

Route being re-routed

This **X** mark could represent an accident, for example. You can also specify a polyline, such as an entire road or a road block. It can also be a polygon to represent a spill of a toxic substance, for instance. Of course, barriers come at a cost because the routing engine will need to compute a new route around the barrier, which results in greater route length, time, or cost.

Barriers are added to a collection, just like stops, but you need to specify the type of geometry. For example, to set point barriers, you need to use `SetPointBarriers`. To use polylines, use `SetPolylineBarriers`. Lastly, to incorporate polygon barriers, use `SetPolygonBarriers`.

Driving directions

Another option you can enable with `RouteParameters` is to return directions. Set `ReturnDirections` to `true` and `RouteTask` will return directions. With `ReturnDirections` you can set the distance units using `DirectionsLengthUnits`. Also, you can specify the language to use with `DirectionsLanguage`. Lastly, you can specify the style of the directions with `DirectionsStyleName`. There are three string options:

- **NA Desktop**: This option is used for turn-by-turn directions suitable for printing
- **NA Navigation**: This option is used for turn-by-turn directions for in-vehicle navigation
- **NA Campus**: This option is used for turn-by-turn walking directions for pedestrian routes

U-turn policy

With `RestrictUTurns`, you can specify whether U-turns are allowed. However, in order to understand these settings, a few key terms need to be understood. A junction is where two polylines meet. An intersection is where three or more polylines intersect. The end of a cul-de-sac is a dead end. There are three options for `UTurnRestriction`, which is the enumeration for `RestrictUTurns`:

Key terms	Description
`Default`	This keyword gives the default setting.
`NoBacktrack`	This keyword means U-turns are prohibited for all junctions, intersections, and dead ends.
`AtDeadEndsOnly`	This keyword means U-turns are allowed at dead ends only.
`AllowBacktrack`	This keyword means U-turns are permitted everywhere.
`AtDeadEndsAndIntersections`	This keyword means U-turns are not allowed at junctions where exactly two adjacent streets meet.

Result options

Another option available on `RouteParameters` is `OutputLines`. With this property, you can specify the complexity of the output routes. The output line can be generalized with `OutputLine` via these enumerations: `Default`, `None`, `Straight`, `TrueShape` and `TrueShapeWithMeasure`. The `Straight` option returns simple point-to-point lines, no geometry is returned with the `None` option, `TrueShape` returns the same shape as the underlying street network, and `TrueShapeWithMeasure` lets `RouteParameters` know to return the accumulated route cost (distance, time, and so on) for the routes. Using some of these options, such as `None` or `Straight`, can improve performance. However, a `LocalRouteTask` class will always return the full geometry payload no matter which option you choose for `OutputLine`.

Geometry options

Another way to generalize polylines is to set the `OutputGeometryPrecision` and `OutputGeometryPrecisionUnits` properties. Generalization is the process of simplifying the geometry so that the essential points (vertices) that make up the polyline are returned. The simplifying process uses a distance to determine which vertices to remove from the polyline. See here:

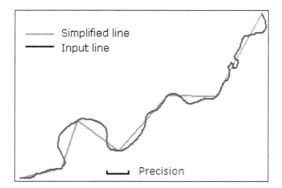

Note here that the input line in black is made of many vertices, but once it is generalized, it produces a much simpler-looking line in red. When you specify the distance as 25 meters, for example, you are telling the generalization algorithm to remove any vertices within that distance. If there are three consecutive vertices and they are within a distance of 25 meters, the middle vertex will be removed. You specify the unit of precision with `OutputGeometryPrecisionUnits`.

Restrictions

In some situations, you are going to want to place constraints on how the routing engine works. For example, you can tell the route to go down one-way streets, which is a requirement for emergency vehicles such as ambulances. To know which restrictions are available for the service, look under `Restrictions` for the service. For example, the San Diego service has the following options: `Avoid Gates`, `Avoid Private Roads`, `Avoid Unpaved Roads`, `Divider Restriction`, `Driving an Automobile`, `Oneway`, and `Through Traffic Prohibited`. To specify these restrictions, add them to a generic list of string objects, like so:

```
List<string> restrictions = new List<string>();
restrictions.Add("Avoid Ferries");
restrictions.Add("Avoid Toll Roads");
routeParameters.RestrictionAttributeNames = restrictions;
```

Time windows

When you need to specify a time frame for your stops, you can specify a time window using `UseTimeWindows` on `RouteParameters`. Just set it to `true`. Then, set the departure time using `StartTime` on `RouteParameters`. Once the route completes, it will return the stops with their `TimeWindowStart` and `TimeWindowEnd`. However, these properties are only available if the analysis layer was created with these options.

Calculating routes

Once parameters are set, the route task can be initiated with `SolveAsync`. The `SolveAsync` method will return a `RouteResult` class, which you can iterate over. Typically, you want the first route returned, but you can also use some other properties, such as the returned route attributes, directions, and any warnings or errors. With this information, you can add more logic to return exactly what the user wants to see on the map, or presented in a side panel.

Example routing app

Let's create an app that illustrates using offline routing for San Francisco. You will need to create an app like we created with `Chapter7a`, and then perform the following steps:

1. Add the following private members:
    ```
    public RelayCommand AddStopsRelayCommand { get; private
        set; }
    GraphicsLayer routeGraphicsLayer = null;
    ```

```
GraphicsLayer stopsGraphicsLayer = null;
LocalRouteTask routeTask;
```

2. In the anonymous MVVM `Messenger` method, add the following lines of code at the end of the method:

```
// Create a new renderer to symbolize the routing polyline
    and apply it to the GraphicsLayer
SimpleLineSymbol polylineRouteSymbol = new
    SimpleLineSymbol();
polylineRouteSymbol.Color =
    System.Windows.Media.Colors.Blue;
polylineRouteSymbol.Style =
    Esri.ArcGISRuntime.Symbology.SimpleLineStyle.Dot;
polylineRouteSymbol.Width = 2;
SimpleRenderer polylineRouteRenderer = new
    SimpleRenderer();
polylineRouteRenderer.Symbol = polylineRouteSymbol;

// Create a new renderer to symbolize the start and end
    points that define the route and apply it to the GraphicsLayer
SimpleMarkerSymbol stopSymbol = new SimpleMarkerSymbol();
stopSymbol.Color = System.Windows.Media.Colors.Green;
stopSymbol.Size = 12;
stopSymbol.Style = SimpleMarkerStyle.Circle;
SimpleRenderer stopRenderer = new .SimpleRenderer();
stopRenderer.Symbol = stopSymbol;

// create the route results graphics layer
this.routeGraphicsLayer = new GraphicsLayer();
this.routeGraphicsLayer.ID = "RouteResults";
this.routeGraphicsLayer.DisplayName = "Routes";
this.routeGraphicsLayer.Renderer = polylineRouteRenderer;
this.routeGraphicsLayer.InitializeAsync();

this.mapView.Map.Layers.Add(this.routeGraphicsLayer);

// create the stops graphics layer
this.stopsGraphicsLayer = new GraphicsLayer();
this.stopsGraphicsLayer.ID = "Stops";
this.stopsGraphicsLayer.DisplayName = "Stops";
this.stopsGraphicsLayer.Renderer = stopRenderer;
this.stopsGraphicsLayer.InitializeAsync();
this.mapView.Map.Layers.Add(this.stopsGraphicsLayer);
```

```
        // Offline routing task.
        routeTask = new
            LocalRouteTask(@"C:\ArcGISRuntimeBook\Data\Networks\
                RuntimeSanFrancisco.geodatabase", "Streets_ND");
```

3. Add the following method:

```
public async void AddStops()
{
        // If the user clicked the SolveRouteButton more than
            once, clear out any existing stops and routes
                graphics.
        routeGraphicsLayer.Graphics.Clear();
        stopsGraphicsLayer.Graphics.Clear();

        try
        {

                // Mouse click 1: setting the start point for the
                    route
                // -------------------------------------------------
                    ---

                // Get the Editor from the MapView.
                Editor startPointEditor = this.mapView.Editor;

                // Get the MapPoint from where the user clicks in
                    the Map
                // Control. This will be the starting point for the
                    route.
                MapPoint startLocationMapPoint = await
                    startPointEditor.RequestPointAsync();

                // Create a new Graphic and set it's geometry to
                    the user
                // clicked MapPoint
                Graphic startPointGraphic = new Graphic();
                startPointGraphic.Geometry = startLocationMapPoint;

                // Add the start point graphic to the stops
                    GraphicsLayer.
                stopsGraphicsLayer.Graphics.Add(startPointGraphic);

                // Mouse click 2: setting the end point for the
                    route
                // -------------------------------------------------
                    ---
```

```
// Get the Editor from the MapView.
Editor endPointEditor = this.mapView.Editor;

// Get the MapPoint from where the user clicks in
   the Map
 // Control. This will be the ending point for the
    route.
MapPoint endLocationMapPoint = await
   startPointEditor.RequestPointAsync();

// Create a new Graphic and set it's geometry to
   the user
// clicked MapPoint
Graphic endPointGraphic = new Graphic();
endPointGraphic.Geometry = endLocationMapPoint;

// Add the start point graphic to the stops
   GraphicsLayer.
stopsGraphicsLayer.Graphics.Add(endPointGraphic);

// Set the arguments for the RouteTask:

// Get the RouteParameters from the RouteTask.
RouteParameters routeParameters = await
   routeTask.GetDefaultParametersAsync();

// Define the settings for the RouteParameters.
   This includes
// setting the SpatialReference,
// ReturnDirections, DirectionsLengthUnit and
   Stops.
routeParameters.OutSpatialReference =
   this.mapView.SpatialReference;
routeParameters.ReturnDirections = false;
routeParameters.DirectionsLengthUnit =
   LinearUnits.Kilometers;

// Define a List of Graphics based upon the user
   start and end
// clicks in the Map Control that will serve as
   input
// parameters for the RouteTask operation.
List<Graphic> graphicsStops = new List<Graphic>();
```

```
graphicsStops.Add(new
    Graphic(startLocationMapPoint));
graphicsStops.Add(new
    Graphic(endLocationMapPoint));

// Set the stops for the Route.
routeParameters.SetStops(graphicsStops);

// Call the asynchronous function to solve the
    RouteTask.
RouteResult routeResult = await
    routeTask.SolveAsync(routeParameters);

// Ensure we got at least one route back.
if (routeResult.Routes.Count > 0)
{

    // Get the first Route from the List of Routes
    Route firstRoute = routeResult.Routes[0];

    // Get the Geometry from the Graphic in the
        first Route.
    Geometry routeGeometry =
        firstRoute.RouteFeature.Geometry;

    // Create a new Graphic based upon the Graphic.
    Graphic routeGraphic = new
        Graphic(routeGeometry);

    // Add the Graphic (a polyline) to the route
        GraphicsLayer.
    this.routeGraphicsLayer.Graphics.Add
        (routeGraphic);

    }
}
catch (System.AggregateException ex)
{
    // There was a problem, display the results to the
        user.
    var innermostExceptions =
        ex.Flatten().InnerExceptions;
    if (innermostExceptions != null &&
        innermostExceptions.Count > 0)
            Messenger.Default.Send<NotificationMessage>
                (new
                NotificationMessage
                    ((innermostExceptions[0]
```

```
                              .Message.ToString())));
        else
            Messenger.Default.Send<NotificationMessage>(new
                NotificationMessage
                    (ex.Message.ToString())));
    }
    catch (System.Exception ex)
    {
        // There was a problem, display the results to the
            user.
        Messenger.Default.Send<NotificationMessage>(new
            NotificationMessage("Error: " +
                ex.Message.ToString())));
    }
    finally
    {
    }
}
```

4. Add the following XAML code to your `MainWindow.xaml` file:

```xml
<StackPanel VerticalAlignment="Center">
    <Button Content="Add Stops" Width="100" Height="40"
        HorizontalAlignment="Left"
        VerticalAlignment="Center"  Command="{Binding
            AddStopsRelayCommand}">
    </Button>
</StackPanel>
```

5. Run the app, and then click on the **Add Stops** button. Click on two locations along the streets as shown below. After you click on the second point, a route should be calculated and displayed on the map along with the stops:

Calculation and display of a route

Refer to the sample app that comes with this book. It's called `Chapter8`.

There was a fair amount of code to just do routing, but it provides us with a good understanding of how to set up routing and execute a simple example. In short, we first added a new `RelayCommand` class called `AddStops`, and some graphics layer private members so that we can store our results. We then added some initialization code in the anonymous method of `Messenger`, along with some symbology for the route and stops. We also instantiated the `RouteTask` class. After that, we added the `AddStops` method.

The `AddStops` method first requests a `MapPoint` class using the `Editor` object of `MapView`. We're going to return to this `Editor` object in the next chapter when we turn our attention to editing. For now, all that is necessary to know is that it allows us to click on the map and return a `MapPoint` class using `RequestPointAsync`. We do this for the beginning `MapPoint` class and the ending `MapPoint` class. We then add these `MapPoint` classes to their `GraphicsLayer` class. Next, we set up the `RouteParameters` class to use the `SpatialReference`, `ReturnDirections`, and `DirectionsLengthUnit` classes of `MapView`,. After this, we add `Graphics` to a generic list and set the `RouteParameters` stops using `SetStops`. Note that `SetStops` can also use features. Finally, we execute the `RouteTask` class using `SolveAsync`. This will return `RouteResults`, which contains a collection of routes that we can use to display in the `GraphicsLayer` route.

We could expand this sample to include both geocoding and routing, by providing the user with the ability to enter addresses and return routes. We could even provide the user with the ability to click and drag the route too, if that was a requirement.

Summary

In this chapter, we've learned about two subjects that are usually linked together: geocoding and routing. With geocoding, we've learned the concepts and how to perform geocoding with online and disconnected data using the locator. We also learned how to find a place. After geocoding was discussed, we learned how to set up network analysis such as routing, and then we learned about the concept of routing. Finally, we created a sample app using local data by modifying one of our previous apps so that it now supports routing. We were also briefly exposed to the `Editor` object, which allows editing, and in the next chapter, we will turn our full attention to this subject.

9

Editing Features

In this chapter, we're going to tackle editing `FeatureLayers`, whether they come from an online or offline data source. The goal of this chapter is to provide you with the necessary know-how to add, update, and delete features. However, before you can accomplish these kinds of tasks, you will need to understand how to prepare the data to support your editing workflows. There will be a discussion on using online feature services, and then turning them into a local Runtime geodatabase. We will also discuss how to synchronize your edits when they are made, while connected or disconnected, and the following topics as well:

- Online versus offline editing
- Preparing for online or offline editing
- Adding features
- Editing features
- Deleting features
- Selecting features
- Committing edits
- Syncing edits

Understanding the editing process

Editing is the process of making changes to features in a feature layer. This includes adding, updating, and deleting features. In a traditional database sense, this is the same as **Create, Retrieve, Update, and Delete** (**CRUD**) operations. Retrieving has already been discussed when we learned about finding and searching data in *Chapter 7, Finding, Querying and Identifying Features,* so here we will focus on `Create`, `Update`, and `Delete`.

When we talk about editing data, we are not only talking about creating, updating, and deleting the field values, but we are also talking about creating, updating, and deleting geometry. As we noted earlier, a feature has both attributes and geometry. You can't create a feature or graphic without geometry and expect it to show up on the map. However, you can create a feature or graphic with just geometry and no attributes. As we saw in *Chapter 5, Geometry and Symbology*, the feature or graphic must also have symbology in order for it to be seen. Once the data is on the map, you can do things such as moving a point or reshaping a polygon feature. Although you can make edits to graphics just like with features, we will only focus on features in this chapter because graphics are always temporary unless you develop your own means to persist them.

No matter whether you're working online or offline, you really need to understand the ArcGIS Runtime editing workflow so that you can get a sense of how it works. Here is a simplified diagram:

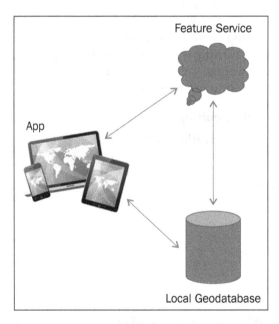

As shown in the diagram, **Feature Service** is required to conduct editing. Also, a **Local Geodatabase** (Runtime geodatabase) is required for storing local edits. Lastly, the **App** can interact with **Feature Service** and/or **Local Geodatabase**. The first thing that needs to be done is that a feature service must be created in ArcMap or Pro. If you plan to do only online editing, a local geodatabase is not required. You can make your edits and they will show up in the feature service as you work with the FeatureLayer class. If you plan to go offline, you can take your edits and put them locally on a device to make edits.

Once you are connected, you can then commit your edits back to the feature service. It's also possible to use both the feature service and Runtime geodatabase at the same time. For example, if you start online and then go offline, your edits will be stored in the Runtime geodatabase, and once you go back online, your edits can be synced with the feature service.

Online and offline editing

If you recall from earlier chapters, we used `FeatureLayers` to view and query data. If using a Runtime geodatabase, we have to access the feature table of `FeatureLayer` with `GeodatabaseFeatureTable` for the Runtime geodatabase. To use a feature service, we get a reference to its `ServiceFeatureTable` object. It's important to understand that you can take edits made to an offline Runtime geodatabase, and then synchronize those edits with the feature service the Runtime geodatabase was created from. However, this synchronization capability must be enabled when creating the feature service. The feature service must be created with ArcGIS Desktop 10.2.2 or higher.

Creating a feature service

Feature services can be created using one of the following approaches:

- ArcGIS Online (`https://www.arcgis.com/home/`)

- ArcGIS Server with or without Portal for ArcGIS

- ArcGIS for Developers (`https://developers.arcgis.com/en/`)

  To create a feature service, go to this site. You will need to create an account if you haven't done so already. Then, you'll need to create a hosted feature service by clicking on **Hosted Data**, and then clicking on **New Feature Service**. After these initial steps, following the remaining steps listed on the site.

To learn more about feature services, navigate to `http://server.arcgis.com/en/server/10.3/publish-services/linux/what-is-a-feature-service-.htm`.

Once you have a feature service, you will also get a REST endpoint that you will use in ArcGIS Runtime. Also, if you plan to make the feature service synchronization enabled, you will need to turn on this capability when you create the service, or you will need to turn the capability on after the service has been created. Lastly, you can symbolize the features using an attribute, and then publish them so that your ArcGIS Runtime app can use them in an editing template. An editing template is one of the extra controls that comes with the runtime toolkit we used in *Chapter 6, Displaying Information*. See the `TemplatePicker` feature at `https://github.com/Esri/arcgis-toolkit-dotnet`.

Downloading the Runtime geodatabase

Once you have a feature service, you will need to then generate a Runtime geodatabase from it so that you can make edits locally, and then synchronize them with the server. There are two ways to accomplish this:

- Using the ArcGIS REST API
- Using the ArcGIS Runtime API

Alternatively, you can create a read-only Runtime geodatabase using one of the following approaches:

- Using ArcGIS Desktop via `ArcMap` to publish a read-only Runtime geodatabase
- Using ArcGIS Desktop via the `Geoprocessing` tool to publish a read-only Runtime geodatabase

Downloading a Runtime geodatabase using the ArcGIS Server REST API

With the ArcGIS Server REST API, the feature service has the ability to create a copy of itself using `CreateReplica`. For more information, navigate to `http://resources.arcgis.com/en/help/rest/apiref/index.html?featureserver.html`. After the edits are made, the feature service can be updated using `SynchornizeReplica`; refer to `http://resources.arcgis.com/en/help/rest/apiref/index.html?featureserver.html`.

If you publish the San Francisco parking meters layer that comes with the book and make sure its synchronization is enabled, you could then go to this feature service's directory and perform the following steps to produce a Runtime geodatabase:

1. Go to the feature service's location. For example, if your service is named `sf_parking_meters`, navigate to `http://localhost/was/rest/services/sf_parking_meters/FeatureServer`.

2. Click on **Create Replica** at the bottom of the page.

3. Give the replica a unique name in **Replica Name**.

4. Specify the layers. In this case, only one layer is available in the feature service, so we specify `0`.

5. Specify the layer's extent in the **Geometry** field, like this:

```
"xmin": -122.7,
"ymin": 37.64,
"xmax": -122.25,
"ymax": 37.9
```

6. Set the spatial reference. With this data, the spatial reference is `4326`.

7. Set **Data Format** to `sqllite`. This is the Runtime geodatabase's format.

8. Click on **Create Replica**. The Runtime geodatabase will be created and the result will be shown below the **Create Replica** button.

9. Once the operation is completed, the URL to the data source will be shown in the output. Copy the URL to another tab in your browser. In the following example, the URL is `http://localhost/was/rest/directories/ arcgisoutput/sf_parking_meters_MapServer/_ags_data{4106EEFA33BB 4A399AE3773F8FC14C3D}.geodatabase`.

Press the *Enter* key in your browser and the Runtime geodatabase will be downloaded from the server. You can now use this Runtime geodatabase to perform edits on. You may want to rename the output file to something more useful for your app:

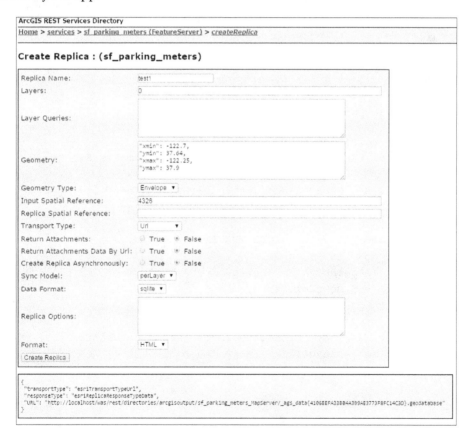

It should also be noted that this entire process can be automated using Python or JavaScript.

Generating a Runtime geodatabase using the ArcGIS Runtime API

The second approach is to use the ArcGIS Runtime API. The general process when using the ArcGIS Runtime API is as follows:

1. Generate the Runtime geodatabase.

2. Make edits to the Runtime geodatabase.

3. Synchronize.

4. Unregister the Runtime geodatabase.

In this section, we'll discuss step 1. In the sections that follow, we'll cover steps 2, 3, and 4.

There are several classes involved in generating a Runtime geodatabase using the GenerateGeodatabaseParameters, GenerateGeodatabaseResult, and GeodatabaseSyncTask APIs. Here is an example of using the objects:

```
var syncTask = new GeodatabaseSyncTask(new Uri(this.URL));

// Get current viewpoints extent from the MapView
var currentViewpoint =
    this.mapView.GetCurrentViewpoint
        (ViewpointType.BoundingGeometry);
var viewpointExtent = currentViewpoint.TargetGeometry.Extent;

var options = new GenerateGeodatabaseParameters(new int[] { 0 },
    viewpointExtent)
{
    GeodatabasePrefixName = GDB_PREFIX,
    ReturnAttachments = false,
    OutSpatialReference = MyMapView.SpatialReference,
    SyncModel = SyncModel.PerLayer
};

var tcs = new TaskCompletionSource<GeodatabaseStatusInfo>();
Action<GeodatabaseStatusInfo, Exception>
    completionAction = (info, ex) =>
{
    if (ex != null)
        tcs.SetException(ex);
        tcs.SetResult(info);
};
```

```
var generationProgress = new Progress<GeodatabaseStatusInfo>();
generationProgress.ProgressChanged += (sndr, sts) => {
    ReportStatus(sts.Status.ToString()); };

// Starting GenerateGeodatabase
var result = await syncTask.GenerateGeodatabaseAsync(options,
    completionAction,
    TimeSpan.FromSeconds(3), generationProgress,
    CancellationToken.None);

// Waiting on geodatabase from server
var statusResult = await tcs.Task;

// Downloading Geodatabase…
var gdbPath = await DownloadGeodatabase(statusResult);

// Create local Feature Layers
```

This code segment creates a `GeodatabaseSyncTask` API with the feature service's URL, gets the extent of the map, and then sets up the parameters for generating the Runtime geodatabase. In this example, only the first layer is used when specifying the `GenerateGeodatabaseParameters` API. Also, `GenerateGeodatabaseParameters` is told to not return attached, and to use a synchronize model called `PerLayer`. The `PerLayer` model indicates that each layer will be synced individually. If the synchronization model is set to `PerGeodatabase`, all layers will be synced once the syncing process starts. A `TaskCompletionSource` instance is created next, which is called when the download finishes. Next, a `Progress` object is created with a `GeodatabaseStatusInfo` object. The `GeodatabaseStatusInfo` object simply lets you know the progress of the download process. After the `Progress` object is created, the `GenerateGeodatabaseAsync` object is started with `GenerateGeodatabaseParameter`, the `TaskCompletionSource` instance, the time span in which it is checked, the progress, and `CancellationToken` of `None`. Next, the task is awaited using `tcs.Task`. Finally, when the task is completed, the Runtime geodatabase is downloaded using this method:

```
private async Task<string>
    DownloadGeodatabase(GeodatabaseStatusInfo statusResult)
{
    var client = new ArcGISHttpClient();
    var gdbStream = client.GetOrPostAsync(statusResult.ResultUri,
        null);
    var gdbFolder = System.IO.Path.GetTempPath();
    var gdbPath = System.IO.Path.Combine(gdbFolder,
        statusResult.GeodatabaseName);
```

```
    if (!System.IO.Directory.Exists(gdbFolder))
        System.IO.Directory.CreateDirectory(gdbFolder);

    await Task.Run(async () =>
    {
        using (var stream = System.IO.File.Create(gdbPath))
        {
            await gdbStream.Result.Content.CopyToAsync(stream);
        }
    });

    return gdbPath;
}
```

The `ArcGISHttpClient` class in `Esri.ArcGISRuntime.Http` is optimized to get data from the ArcGIS Server REST API. Once it is instantiated, you call `GetOrPostAsync` with the URI of `GeodatabaseStatusInfo` to the service and it will return an `HttpResponseMessage` message stream (`HttpResponseMessage`). With this stream, you can create a geodatabase file and fill it with the content of the stream's contents. As a side note, `Esri.ArcGISRuntime.Http` allows you to write your own interfaces for interacting with ArcGIS Server.

Generating a Runtime geodatabase from ArcGIS Desktop

To create a read-only Runtime geodatabase from ArcGIS Desktop, you will need perform the following steps:

1. Add the layer to `ArcMap` and set up required symbology.
2. In `ArcMap`, click on **Customize | ArcMap Options**.
3. In the **ArcMap Options** dialog box, click on **Sharing**.
4. Check the **Enable ArcGIS Runtime Tools** checkbox and click on **OK**.
5. Click on **File | Share As | ArcGIS Runtime Content**.
6. Specify the `Output` folder below **Save ArcGIS Runtime content to**.
7. Click on **Analyze** to find any errors, and then resolve them.
8. Click on **Share**. This will create the Runtime geodatabase file with the `.geodatabase` file extension.

This workflow is easy to execute and provides a simple way to generate a Runtime geodatabase if your users don't need to edit the data.

Generating a Runtime geodatabase using a Geoprocessing tool

We will now generate a Runtime geodatabase using a `Geoprocessing` tool by performing the following steps:

1. Create the map document using `ArcMap` with the feature layers.

2. In **ArcToolbox,** click on **Data Management Tools | Package | Create Runtime Content.**

3. Fill in the tool's parameters and click on **OK:**

Once again, this will create a read-only Runtime geodatabase.

Editing features

Now that we've covered the basic concepts of how to create a Runtime geodatabase, let's discuss how this actually works in the app. In this section, we'll cover how to create a feature table; adding, updating, and deleting features; and selecting and committing your edits.

The basic workflow includes the following steps:

1. Create the feature table. If you are using a feature service, you will access the service via its feature table. If you're using an offline service, you will access it from the Runtime geodatabase that was sync enabled.

2. Create a `FeatureLayer` class from the feature table and add it to the map.

3. Edit the layer (add, update, or delete features).

4. Commit your edits. If you're using a feature service, apply the edits. If you're using an offline service, synchronize the edits back to the feature service.

Creating a FeatureLayer class

You've already seen code for step 1 in this list of tasks. For example, when we worked with the parking meters layer, we opened up the Runtime geodatabase using this code:

```
private async void CreateFeatureLayers()
{
    var gdb = await Geodatabase.OpenAsync(this.GDB);

    foreach (var table in gdb.FeatureTables)
    {
        var flayer = new FeatureLayer()
        {
            ID = table.Name,
            DisplayName = "Parking Meters",
            FeatureTable = table
        };
        this.mapView.Map.Layers.Add(flayer);
    }
}
```

This code opens the offline geodatabase, iterates over all the `FeatureTables` property, creates `FeatureLayers`, and finally adds them to the map. The first class to note is `Geodatabase`. This is a simple class but it does something very important. It allows us to open a Runtime geodatabase using a path. It has a property on it called `FeatureTables` that allows us to get all `FeatureTables` in the Runtime geodatabase. With the `FeatureTable` class, we can create layers and add them to the map.

Let's also take a look at an online service:

```
public async void CreateOnlineFeatureService()
{
    // create a new geodatabase feature service table; point to a
    // feature service layer end point
    var table = new ServiceFeatureTable();
    table.ServiceUri =
        "http://sampleserver6.arcgisonline.com/arcgis/rest/
            services/Military/FeatureServer/2";

    // initialize the feature table; use the map view's spatial
    // reference for the optional "sr_override" argument
    await table.InitializeAsync(this.mapView.SpatialReference);

    // if the table was initialized successfully, create a new
    // feature layer for the table and add it to the map
    if (table.IsInitialized)
    {
        var lyr = new FeatureLayer
        {
            ID = "Units",
            DisplayName = "Units",
            FeatureTable = table
        };

        this.mapView.Map.Layers.Add(lyr);
    }
}
```

Instead of having to open the `Geodatabase` class, as with offline data, we simply created the `ServiceFeatureTable` object directly, initialized it, and then created a `FeatureLayer` class. The end result is the same: we now have a feature service we can make edits to. If you look at the second layer in this feature service, you will note that it comes with `UniqueValueRender` that uses the `MIL-STD-2525C` symbology already established. As a result, a user could start making edits to this layer immediately and it will use this renderer when features are added to the `FeatureLayer` class.

By default, a feature service will only retrieve 1,000 records at a time, so you may not see all of the features in the feature service. You can alter this by changing `MaxRecordCount` on `FeatureService` to an acceptable number. Keep in mind, however, that the more features you pull across the Web, the slower your app will be. One very important step you can make when creating the feature service is to set a scale dependency on it so that it only renders the layer at the intended scale of its use. If the layer doesn't need to be seen beyond 1:100,000, make sure it doesn't show up beyond that scale. This will not only make the map more readable, it will also improve performance.

Editing the FeatureLayer class

In this section, we'll discuss how to add, edit, delete, and select features in a `FeatureLayer` class. Before we go over these topics, let's turn our attention to the `Editor` object.

Editor

To edit features in ArcGIS Runtime, you must use the `Editor` class. This class is of such importance that it is directly available on the `MapView` object. You can access this object like this:

```
Editor editor = this.mapView.Editor;
```

With the `Editor` object, you can request a point using `RequestPointAsync`, request other geometry types (`Point`, `Polyline`, `Polygon`, `Enveloper`, `Rectangle`, `Freehand`, `Arrow`, `Triangle`, `Ellipse`, `Circle`, or `LineSegment`) using `RequestShapeAsync`, and edit geometry with `EditGeometryAsync`. All three of these methods can be called from an MVVM Light `RelayCommand`. See here for an example:

```
// wait for user to draw the shape
var geometry = await
    this.mapView.Editor.RequestShapeAsync(DrawShape.Polygon,
        symbol);
```

Once the method is invoked, these methods expect you to do something on the map, such as click on a single point or click on multiple locations for a polygon. The `RequestShapeAsync` method expects an enumeration of the type of geometry to draw. Also, note that we provided a symbol so that we can see the polygon after we've added it. Lastly, in order to draw a point, you simply have to click on the map once. To add a polyline or polygon, you have to click on the map once for each point that makes it up, and then for the final point of the polyline or polygon, you need to double-click on the map. If you use the `LineSegment` option, you will need to click-drag-release (touch-drag-release) on the map to create a line that only has two vertices.

Editing the configuration

Although the `Editor` object comes out of the box ready for you to work with the map, you can also configure how it works. To configure the Editor object, use `Esri.ArcGISRuntime.Controls.EditorConfiguration`. This object allows you configure the editor to support certain operations. For example, you can tell the `Editor` object to allow for adding, deleting, or moving vertices (points) on a multipart feature such as a polyline using `AllowAddVertex`, `AllowDeleteVertex`, and `AllowMoveVertex`, respectively. The following screenshot shows an example of vertices along a polyline geometry:

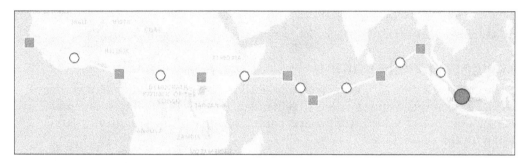

As you can see in the preceding screenshot, the vertices have been given a unique symbol. The **Vertex** symbol is set to a solid blue square. The **Polyline** symbol is yellow. The **MidVertex** symbol is a white circle. The **SelectedVertex** symbol is a red circle.

You can also allow the user to rotate geometry or not using `AllowRotateGeometry`. Here's an example of turning on the ability to add vertices and setting the symbology for the `Editor` object:

```
// Create a new EditorConfiguration which allows for the changing
   of
// the various MarkerSymbols (SelectedVertexSymbol, VertexSymbol,
   and
// MidVertexSymbol).
Esri.ArcGISRuntime.Controls.EditorConfiguration
   editorConfiguration =
   new Esri.ArcGISRuntime.Controls.EditorConfiguration();

editorConfiguration.AllowAddVertex = true;

// Set the EditorConfiguration.MidVertexSymbol
SimpleMarkerSymbol midVertexSymbol = simpleMarkerSymbol;
editorConfiguration.MidVertexSymbol = midVertexSymbol;
```

```
// Set the EditorConfiguration.VertexSymbol to the
    SimpleMarkerSymbol
SimpleMarkerSymbol vertexSymbol = simpleMarkerSymbol;
editorConfiguration.VertexSymbol = vertexSymbol;

// Set the EditorConfiguration.SelectedVertexSymbol
Esri.ArcGISRuntime.Symbology.SimpleMarkerSymbol selectedVertexSymbol =
simpleMarkerSymbol;
editorConfiguration.SelectedVertexSymbol = selectedVertexSymbol;

// Set the Editor.EditorConfiguration to the custom
    editorConfiguration.
this.mapView.Editor.EditorConfiguration = editorConfiguration;
```

Navigating while editing

While creating a new geometry, you may need to navigate with the map or vice versa. For example, you may need to pan or zoom while digitizing a polygon. The Editor object has a few properties and commands that allow you to effectively deal with this scenario:

Name	Usage
IsActive	This property indicates whether you are in the middle of an edit or sketch.
IsSuspended	This property indicates whether the Editor object has been prevented from responding to a platform event property.
Cancel	This command prevents the Editor object from returning geometry.
Complete	This command allows the Editor object to return a geometry before it is deactivated.

The IsActive property is an indispensable property when trying to edit. For example, if you have an event handler such as MapViewTapped that performs similar to Identify, or shows an overlay, it will be necessary to cancel this event; otherwise, an error can occur when doing something like adding a polyline using RequestShapeAsync:

```
private async void AssociatedObject_MapViewTapped(object sender,
    MapViewInputEventArgs e)
{

    MapView mapView = sender as MapView;

    if (mapView.Editor.IsActive)
```

```
        return;

    // do something when tapped

}
```

Edit progress

While editing or measuring, RequestShapeAync can be used to track sketch progress. By using IProgress<GeometryEditStatus>, you can track progress such as the length of a polyline as you click from one point to another. Capturing the progress requires a two-step process:

1. Declare an instance of Progress<GeometryEditStatus>. This object accepts an Action delegate, which you can use to respond to the sketch.

2. Pass the Progress instance as a parameter to RequestShapeAsync:

   ```
   // Track progress of the sketch
   var progress = new
       Progress<GeometryEditStatus>(OnStatusUpdated);

   // create a Simple Fill Symbol
   SimpleFillSymbol sfs = new SimpleFillSymbol();
   sfs.Style = SimpleFillStyle.Null;
   sfs.Outline = new SimpleLineSymbol()
   {
       Color = System.Windows.Media.Colors.Blue,
       Style = SimpleLineStyle.Dash
   };

   // Obtain the rectangle using a SimpleFillSymbol and track
       the progress
   Geometry geometry = await
       myEditor.RequestShapeAsync(DrawShape.Rectangle, sfs,
           progress);
   ```

If you use this code, it will now show a rectangle with a dashed blue line. The GeometryEditStatus delegate (OnStatusUpdated) captures the current state of the sketch using the NewGeometry property, as shown here:

```
private void OnStatusUpdated(GeometryEditStatus status)
{
    var polyline = status.NewGeometry as Polyline;
}
```

With the polyline, you can report the length of it as you move the mouse. Also, with GeometryEngine, you can convert the length to its geodesic equivalent.

Adding features

To add a new feature to the map, you'd use code similar to the following code, which adds a parking meter to the map:

```
public async void AddPoint()
{
    // get a point from the user
    var mapPoint = await this.mapView.Editor.RequestPointAsync();
    // open the local geodatabase
    var gdb = await Geodatabase.OpenAsync(this.GDB);

    // open the table and create a new feature using its schema
    FeatureTable gdbTable;

    foreach (FeatureTable table in gdb.FeatureTables)
    {
        gdbTable = table;

        var newFeature = new GeodatabaseFeature(gdbTable.Schema);

        // assign the point from the user as the feature's
            geometry
        newFeature.Geometry = mapPoint;

        // get the attributes from the feature
        // (a Dictionary<string, object>) and set some values
        var attributes = newFeature.Attributes;
        attributes["POST_ID"] = "123456";

        // add the new feature to the table, the OID of the new
        // feature is returned
        var recNo = await gdbTable.AddAsync(newFeature);

        break;
    }


    await this.mapView.SetViewAsync(mapPoint, this.mapView.Scale +
        1);
}
```

This code first requests `MapPoint` asynchronously from the map, opens the Runtime geodatabase, iterates over the `FeatureTables` property so that it can create `GeodatabaseFeature` using the same attributes of the layer's attributes (`gdbTable.Schema`), sets the geometry of `newFeature` with `Map`, sets an attribute, adds the feature to the table, and finally zooms to the newly added feature. The two most important things happening here are that the new feature is created in the following ways:

- Using the geometry of where the user clicks. It's not a feature unless it has geometry.
- Using the schema of the table. This is necessary because you want the new feature to have the same attribute scheme as the `FeatureLayer` class.

The other important thing that happens here is that the attribute is set using the syntax of a generic dictionary. In a professional app, we would prompt the user for the values to enter based on the table's fields, instead of hardcoding the value as we did in this example.

Selecting features

Typically, before you update or delete a feature, you need to have your user select the feature so that they can perform some operation. Selection is the process of placing a feature into a selected state so that you can do something with it, such as changing its attributes or moving it. To select features, you will need to perform one of the following tasks:

- Use a hit test, which allows the user to click on a feature and return the ID, which can then be passed to `FeatureLayer.SelectFeatures`.
- Query or find them and return their IDs.

Here is an example that selects an individual parking meter on the map:

```
private async void AssociatedObject_MapViewTapped(object sender,
    MapViewInputEventArgs e)
{
    MapView mapView = sender as MapView;

    // get the location tapped on the map
    var mapPoint = e.Location;

    // Get the second layer
    FeatureLayer featureLayer =
        (FeatureLayer)mapView.Map.Layers[1];
```

```
        // Get possible features
        var foundFeatures = await featureLayer.HitTestAsync(mapView,
            new Rect(e.Position, new Size(10, 10)), 1);

        featureLayer.SelectFeatures(foundFeatures);
    }
```

The HitTestAsync method returns the list of IDs of the features found when the user taps on the map. It only returns the IDs of the features in the second FeatureLayer class. Then, these IDs are passed to SelectFeatures and the features are highlighted on the map like so:

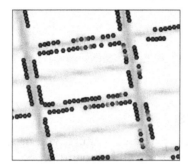

This approach is fine if you want to select features individually, but if you want to drag a rectangle on the map and select objects in bulk, you would use code similar to the following code in Map:

```
public async void SelectByRectangle()
{
    try
    {
        FeatureLayer featureLayer =
            (FeatureLayer)this.mapView.Map.Layers[1];

        featureLayer.ClearSelection();// Clear the selection

        // Get the Editor associated with the MapView. The Editor
        // enables drawing and editing graphic objects.
        Editor editor = this.mapView.Editor;

        // Get the Envelope that the user draws on the Map.
          Execution
        // of the code stops here until the user is done drawing
          the
        // rectangle.
```

```
Geometry geometry = await
    editor.RequestShapeAsync(DrawShape.Rectangle);
Envelope envelope = geometry.Extent;

if (envelope != null)
{
    // Get the lower-left MapPoint (real world
    //   coordinates)
    // from the Envelope the user drew on the Map and then
    // translate it into
    // a Microsoft Point object.
    MapPoint mapPoint1 = new
        MapPoint(envelope.Extent.XMin,
        envelope.YMin);
    System.Windows.Point windowsPoint1 =
        this.mapView.LocationToScreen(mapPoint1);

    // Get the upper-right MapPoint (real world
    //   coordinates)
    // from the Envelope the user drew on the Map and then
    // translate it into
    // a Microsoft Point object.
    MapPoint mapPoint2 = new
        MapPoint(envelope.Extent.XMax,
        envelope.YMax);
    System.Windows.Point windowsPoint2 =
        this.mapView.LocationToScreen(mapPoint2);

    // Create a Windows Rectangle from the Windows Point
    // objects.
    System.Windows.Rect windowsRect = new
        System.Windows.Rect(windowsPoint1, windowsPoint2);

    // Get the FeatureTable from the FeatureLayer.
    FeatureTable featureTable =
        featureLayer.FeatureTable;

    // Get the number of records in the FeatureTable.
    long count = featureTable.RowCount;

    // Use the FeatureLayer.HitTestAsync Method to
    //   retrieve
    // the FeatureLayer IDs that are within or cross the
    // envelope that was drawn on the Map by the user. It
    //   is
```

```
                    // important to note that by passing in the variable
                    // count (which is the maximum number of features in
                        the
                    // FeatureLayer), you are able to select up to the
                        number
                    // of features in the FeatureLayer. If you were to
                        leave
                    // off this optional last parameter then the
                        HitTestAsync
                    // would only return one record!
                    long[] featureLayerRowIDs = await
                        featureLayer.HitTestAsync(this.mapView,
                            windowsRect,
                          System.Convert.ToInt32(count));

                    if (featureLayerRowIDs.Length > 0)
                    {
                        // We have at least one record in the FeatureLayer
                        // selected.

                        // Cause the features in the FeatureLayer to
                        // highlight (cyan) in the Map.
                        featureLayer.SelectFeatures(featureLayerRowIDs);

                    }
                }
            }
        catch (System.Threading.Tasks.TaskCanceledException)
        {
            // This exception occurred because the user has already
            // clicked the button but has not drawn a rectangle on the
            // Map yet.
              Messenger.Default.Send<NotificationMessage>(new
              NotificationMessage(
                  "Drag a rectangle across the map to select some
                  features."));      }
        catch (System.Exception ex)
        {
            // We had some kind of issue. Display to the user so it
                can

    // be corrected.
    Messenger.Default.Send<NotificationMessage>(new
        NotificationMessage(ex.Message));
        }
    }
```

As shown in the left-hand screenshot, the user will drag a box, and then once completed, all parking meters within the box are now in a selected state in the right-hand image:

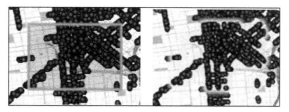

Dragging and selection

Updating features

To update features, use the UpdateAsync method on GeodatabaseFeatureTable. The following example illustrates how to do this:

```
public async void UpdateSelectedFeaturesAsync()
{
    // create a list of the record IDs to update
    var idList = new List<long> { this.lastAddedOid };

    // query the table for the features with the specified IDs
    var updateFeatures = await gdbTable.QueryAsync(idList);

    // get the first GeodatabaseFeature (should be one or zero)
    var feature = updateFeatures.FirstOrDefault();
    if (feature == null) { return; }

    // update the feature's geometry with a new point
    feature.Geometry = newPoint;

    // update one of the attribute values
    feature.Attributes["comments"] = "point updated to last known
        location";

    // commit the updated feature to the GeodatabaseTable
    await gdbTable.UpdateAsync(feature);

}
```

In this example, we have a private member variable that keeps a list of features added to the map. We then query for the features, get the first one, update its geometry and attributes, and finally update the feature using UpdateAsync.

Deleting features

Deleting features from `GeodatabaseFeatureTable` is easy to accomplish with `DeleteAsync`. This method has several overloads, which allow you to specify a single feature ID or a list of features using `IEnumerable<long>`. If you graphically select features using a rectangle, as we did earlier, you can return the selected feature IDs using `FeatureLayer.SelectedFeatureIDs`, and then call `DeleteAsync` with the IDs.

Committing edits

Once you've made edits, you now need to allow your users to commit their changes to the feature service or Runtime geodatabase. Whether the app is fully connected or not determines the commit process you will follow. In situations where two users are working on the same feature, the user that edits the feature last will have their edits seen. In other words, it's possible to overwrite each other's work.

Online

While fully connected using a feature service, you can apply edits and other users will see your edits immediately. If you call `ApplyEditsAsync` on the feature service's `ServiceFeatureTable`, your edits will be sent to the server. The `ApplyEditsAsync` method has a parameter called `rollbackOnFailure`. When this parameter is set to `true`, all features must successfully make it to the server, or the syncing process will fail. If it is set to `false`, individual features can make it to the server without the process failing. You can use `AddedFeaturesCount`, `UpdatedFeaturesCount`, and `DeletedFeatureCount` to get counts on the number of features added, updated, and deleted, respectively.

The `ApplyEditsAsync` method returns a `FeatureEditResult` object that contains a set of properties about individual features, which were attempted to be sent to the server. The results include the following properties: `AddResults`, `UpdateResults`, and `DeleteResults`. Each of these properties returns a read-only list of `FeatureEditResultItems`, which provides the Object ID, Global ID, success flag, and error details (if any occurred) of the edited features. The global ID is a unique GUID of the feature in which the edit was attempted. Here's some example code:

```
var table = (ArcGISFeatureTable)layer.FeatureTable;
string message = null;
try
{
    if (!table.HasEdits)
        return;
    if (table is ServiceFeatureTable)
    {
```

```
            var serviceTable = (ServiceFeatureTable)table;
            // Pushes accumulated edits back to the server.
            var saveResult = await serviceTable.ApplyEditsAsync();
            if (saveResult != null && saveResult.AddResults != null
                && saveResult.AddResults.All(r => r.Error == null &&
                r.Success))
                message = string.Format("Saved {0} features",
                        saveResult.AddResults.Count);
        }
    }
    catch (Exception ex)
    {
        message = ex.Message;
    }
```

In this code sample, we first cast the layer's `FeatureTable` to `ArcGISFeatureTable`, check to see whether any edits were made, get the service's table, and then apply the edits. With the results, we check to see there are any `AddResults` property that contain the object ID, global ID, success flag, and any error details. If errors occur during the commit process, you can deal with them accordingly. You can find a complete list of error codes for feature services at `http://resources.arcgis.com/en/help/arcgis-rest-api/#/Apply_Edits_Feature_Service/02r3000000nv000000/`.

Offline

As we showed earlier in this chapter, in the *Generating a Runtime geodatabase using the ArcGIS Runtime API* section, you will need to generate a Runtime geodatabase in order to work offline. When users are ready to commit their edits, you will need to provide them with a means to call `GeodatabaseSyncTask`, which will take their edits to the Runtime geodatabase and send them to the feature service. This synchronization process can send and receive edits. For example, if you have 20 users making edits to a Runtime geodatabase and the fourth user makes edits and then syncs them, all other users will see these edits once they sync their edits (even if they didn't make any edits themselves). As with working exclusively online, the last person to edit a feature will have the last say on where that feature is located and its attributes.

To use `GeodatabaseSyncTask`, you must perform the following steps:

1. Create a `System.IProgress` object so that you can track the progress of the sync process.

2. Set up a `callback` method to report on progress, so that users know if the process succeeds or fails.

3. Create or obtain the `sync` parameters.

4. Create `GeodatabaseSyncTask`.

5. Call the `SyncGeodatabaseAsync` method on `GeodatabaseSyncTask` using the `callback` and `sync` parameters in the earlier steps.

Let's look at an example of syncing with `GeodatabaseSyncTask`:

```
// Synchronizing Local and Online data...
var tcs = new TaskCompletionSource<GeodatabaseStatusInfo>();
Action<GeodatabaseStatusInfo, Exception> completionAction = (info,
    ex) =>
{
    if (ex != null)
        tcs.SetException(ex);
        tcs.SetResult(info);
};

var syncProgress = new Progress<GeodatabaseStatusInfo>();
syncProgress.ProgressChanged += (sndr, sts) => {
    this.secondaryStatus =
    sts.Status.ToString(); };

var syncTask = new GeodatabaseSyncTask(new
    Uri(this.featureService));
var gdbTable = layer.FeatureTable as GeodatabaseFeatureTable;
await syncTask.SyncGeodatabaseAsync(gdbTable.Geodatabase,
    completionAction,
    null,
    TimeSpan.FromSeconds(3),
    syncProgress,
    CancellationToken.None);

await tcs.Task;
```

If you recall from learning .NET, a `TaskCompletionSource` instance is a factory method for creating delayed tasks, ones that won't actually be scheduled until some user-supplied timeout has occurred. In this example, the time is 3 seconds, which is how often `GeodatabaseSyncTask` will check the server when it executes. A `Progress` object is also created with `GeodatabaseStatusInfo`. The `Progress` object is set to a property called `secondaryStatus`. Then, the `GeodatabaseSyncTask` API starts with the supplied URL. Next, `SyncGeodatabaseAsync` is awaited with all of the supplied parameters.

The `GeodatabaseStatusInfo` object can also report `GeodatabaseName`, `LastUpdateTime`, `ResultUri`, `Status`, and `SubmissionTime`. The `ResultUri` property is the Runtime geodatabase name we saw earlier when we downloaded directly from the feature service. See here for an example: `http://localhost/was/rest/directories/arcgisoutput/sf_parking_meters_MapServer/_ags_data{4106EEFA33BB4A399AE3773F8FC14C3D}.geodatabase`. Status includes the following values during the download process: `Unknown`, `Pending`, `InProgress`, `ExportAttachments`, `ExportChanges`, `ExportingData`, `ExportingSnapshot`, `ImportChanges`, `ImportAttachments`, `ProvisioningGeodatabase`, `UnregisteringGeodatabase`, `Completed`, `CompletedWithErrors`, and `Failed`. Two other very useful properties of `GeodatabaseStatusInfo` are `LastUpdatedTime` and `SubmissionTime`. These two properties allow you to keep track of when users submit edits, and can be used to keep metrics on your app's usage. This can be very helpful in sizing your ArcGIS Server and its configuration.

Once this task completes, all of the edits are sent to the feature service. You should refresh the map when completed so that the user can see that their layer has been sent back to the server. You will also want the user to see the feature service as it stands after your commit, so that they get a sense of other edits made (if any).

Once the syncing process has completed, you can unregister the Runtime geodatabase using `UnregisterGeodatabaseAsync`. This method in effect tells ArcGIS Server that it is no longer being used by a particular user's Runtime geodatabase.

Attachments

With feature services, it's possible to attach images, text files, audio, video, and so on, to individual features, as shown here in ArcGIS Online:

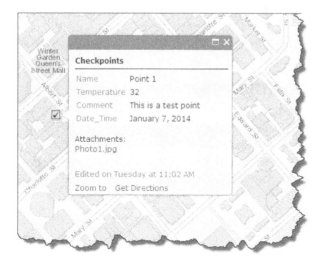

With attachments, your users can take photos with their devices, and then attach them to individual features. For example, if your user is a biologist, they could take snapshots or videos of birds and attach them to the location where the species was sighted. Or, with another example, a utility field crew member could capture the location of transformers and include a photo of them.

To enable attachments, the feature service must have this capability turned on. If the data was created with ArcGIS Desktop, it can be enabled using the **Catalog** window or with **ArcCatalog**, by following the instructions available at `https://desktop.arcgis.com/en/desktop/latest/manage-data/editing-attributes/enabling-attachments-on-a-feature-class.htm`. If you have a hosted feature service with attachments, you can enable them by simply clicking on the layer in ArcGIS Online, as shown here:

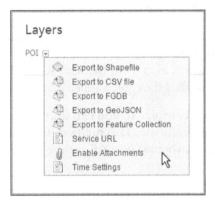

The following file types are supported: 7Z, AIF, AVI, BMP, DOC, DOCX, DOT, ECW, EMF, EPA, GIF, GML, GTAR, GZ, IMG, J2K, JP2, JPC, JPE, JPEG, JPF, JPG, JSON, MDB, MID, MOV, MP2, MP3, MP4, MPA, MPE, MPEG, MPG, MPV2, PDF, PNG, PPT, PPTX, PS, PSD, QT, RA, RAM, RAW, RMI, SID, TAR, TGZ, TIF, TIFF, TXT, VRML, WAV, WMA, WMF, WPS, XLS, XLSX, XLT, XML, and ZIP.

Obtaining attachments

It's important to understand that attachments are by default not included when you download a Runtime geodatabase from a feature service. To include attachments, make sure to set `GenerateGeodatabaseParameters.ReturnAttachment` to `true`. You can also get a feature table of `FeatureLayer` and check whether it has attachments using `HasAttachments`. See here:

```
// get all feature layers from the map
var featureLayers = mapView.Map.Layers.OfType<FeatureLayer>();
// loop thru all feature layers, see if they support attachments
foreach (var lyr in featureLayers)
```

```
{
    var table = lyr.FeatureTable as ArcGISFeatureTable;
    if (table == null) { continue; }

    var hasAttachments = table.HasAttachments;
    // ... do something with attachments ...
}
```

Once you know that a feature table supports attachments, you can find information about the features in question using QueryAttachmentAsync on ArcGISFeatureTable. First, you will need to either iterate over the features or select one of them so that you can obtain its ID. An ID is passed into QueryAttachmenAsync. See here:

```
// Performs hit test on layer to select feature.
var features = await layer.HitTestAsync(MyMapView, e.Position);
if (features == null || !features.Any())
    return;
var featureID = features.FirstOrDefault();
layer.SelectFeatures(new long[] { featureID });

var tableToQuery = (ArcGISFeatureTable)layer.FeatureTable;
var attachments = await
    tableToQuery.QueryAttachmentsAsync(featureID);
```

This code segment could be placed in a MapViewTapped event handler. It simply gets the first feature found in this Hit Test, gets the layer's table, and then returns any attachments. Once you have the attachments, you will need to do something with this, such as show them in a list so that users can view or edit them. See here:

```
// check if any were found
if (attachments.Infos != null && attachments.Infos.Any())
{
    foreach (var info in attachments.Infos)
    {
        // ... process each attachment result ...
    }
}
```

Attachment information

If attachments are found, you can find information about them using the `AttachmentInfo` object. This class provides properties such as the filename, ID, content type, and size in bytes. As returning the attachment data can be an expensive and time-consuming operation, the attachment is not by default included with `AttachmentInfo`.

Attachment data

Once you have the `AttachmentInfo` object, you can iterate over the attachments and do something with them, such as display an image or play a sound. See here:

```
foreach (var info in attachments.Infos)
{
    // find image attachments: "image/jpeg", "image/png", etc.
    if (info.ContentType.Contains(@"image/"))
    {
        // do something with the image
    }
}
```

Editing attachments

Now that you have attachments, you can further evaluate the feature service to determine if it supports adding (`CanAddAttachment`), updating (`CanUpdateAttachment`), and deleting (`CanDeleteAttachment`). Once you know what the feature service is set up to do, you can apply an attachment by simply calling `ApplyAttachmentEditsAsync` on the `ServiceFeatureTable` object. If your workflow requires the ability to revert changes, you can call `ClearEdits`. This will discard any edits made on the table. This will not only remove the attachments, but will also remove any feature edits.

Adding an attachment

To add an attachment, call `AddAttachmentAsync` on `ArcGISFeatureTable`. This method requires a .NET stream representing the data, a filename, and a target feature ID. This edit operation will be in-memory until calling `ApplyAttachmentEditsAsync`. A new feature can be added provided that the following conditions are met:

- The feature service supports attachments.
- The size of the file must be less than the size ArcGIS Server is set up to handle. The default size is 2 GB.
- It must be one of the supported files listed earlier.

Here is an example of adding an attachment:

```
var table = (ArcGISFeatureTable)layer.FeatureTable;
var file = GetFile();
if (file == null) return;

AttachmentResult addResult = null;
using (var stream = file.OpenRead())
{
    addResult = await table.AddAttachmentAsync(featureID, stream,
        file.Name);
}
// save edits
```

We obtain the layer's feature table, read in a file as a stream, and then call `AddAttachmentAsync`.

Deleting an attachment

Deleting attachments is very similar to adding. Just pass in some feature IDs:

```
DeleteAttachmentResult deleteResult = null;
deleteResult = await table.DeleteAttachmentsAsync(featureID,
    new long[] { info.ID });
if (deleteResult != null && deleteResult.Results != null &&
    deleteResult.Results.Count > 0)
{
    var result = deleteResult.Results[0];
    if (result.Error != null)
    message = string.Format("Delete attachment [{0}] of feature
        [{1}]
            failed.\n {2}", info.ID, featureID,
                result.Error.Message);
    // save edits
}
```

We pass in the ID of the attachment with `Info.ID`, along with the feature ID to `DeleteAttachmentsAsync`. After that, we check the result to see if it worked correctly, and then capture a message to present to the user.

Updating attachments

Lastly, updating an attachment is very similar to deleting; here, you pass in the attachment IDs:

```
AttachmentResult updateResult = null;
using (var stream = file.OpenRead())
{
    updateResult = await table.UpdateAttachmentAsync(featureID,
        info.ID, stream, file.Name);
}
if (updateResult != null)
{
    if (updateResult.Error != null)
        message = string.Format("Update on attachment [{0}]
        of feature [{1}] failed.\n {2}", info.ID, featureID,
            updateResult.Error.Message);
    // save edits
}
```

Summary

In this chapter, we've covered nearly all aspects of editing with ArcGIS Runtime. We discussed the differences between editing online and offline, and how, by combining both, to create a seamless editing experience for users as they go from a connected to a disconnected setting. We've also discussed how to set up a Runtime geodatabase for editing from a feature service too. Then, we covered how to create a feature layer from a feature service or from a Runtime geodatabase. After we got access to a feature layer, we discussed the Editor object and its significance in the editing process, along with its configuration. Once everything is set up, we covered the editing operations: adding, selecting, updating, deleting, and committing our edits. Lastly, we discussed how attachments are created, and then edited, so that we can provide our users with an interesting mapping experience.

Now that we've covered most of the basics of using ArcGIS Runtime, we will turn our attention to what makes GIS awesome: spatial analysis.

10
Spatial Analysis

In this chapter, we're going to learn about spatial analysis with ArcGIS Runtime. As with other parts of ArcGIS Runtime, we really need to understand how spatial analysis is set up and executed with ArcGIS Desktop/Pro and ArcGIS Server. As a result, we will first learn about spatial analysis within the context of geoprocessing. Geoprocessing is the workhorse of doing spatial analysis with Esri's technology. Geoprocessing is very similar to how you write code, in that you specify some input data, do some work on that input data, and then produce the desired output. The big difference is that you use tools that come with ArcGIS Desktop or Pro. If you recall, in *Chapter 5*, *Geometry and Symbology*, we did some analysis when we created a buffer using the `GeometryEngine` tool. This same tool is also available in ArcGIS Desktop or Pro, along with dozens of other tools. In this chapter, we're going to learn how to use these tools, and how to specify their input, output, and other parameters from an ArcGIS Runtime app that goes well beyond what's available in the `GeometryEngine` tool. In summary, we're going to cover the following topics:

- Introduction to spatial analysis
- Introduction to geoprocessing
- Preparing for geoprocessing
- Using geoprocessing in runtime
- Online geoprocessing
- Offline geoprocessing

Introducing spatial analysis

Spatial analysis is a broad term that can mean many different things, depending on the kind of study to be undertaken, the tools to be used, and the methods of performing the analysis, and is even subject to the dynamics of the individuals involved in the analysis. In this section, we will look broadly at the kinds of analysis that are possible, so that you have some context as to what is possible with the ArcGIS platform. Spatial analysis can be divided into these five broad categories:

- Point patterns
- Surface analysis
- Areal data
- Interactivity
- Networks

Point pattern analysis is the evaluation of the pattern or distribution of points in space. With ArcGIS, you can analyze point data using average nearest neighbor, central feature, mean center, and so on. For surface analysis, you can create surface models, and then analyze them using tools such as LOS, slope surfaces, viewsheds, and contours. With areal data (polygons), you can perform hotspot analysis, spatial autocorrelation, grouping analysis, and so on. When it comes to modeling interactivity, you can use tools in ArcGIS that allow you to do gravity modeling, location-allocation, and so on. Lastly, with Esri's technology you can analyze networks, such as finding the shortest path, generating drive-time polygons, origin-destination matrices, and many other examples.

ArcGIS provides the ability to perform all of these kinds of analysis using a variety of tools. For example, here the areas in green are visible from the top of the tallest building. Areas in red are not visible:

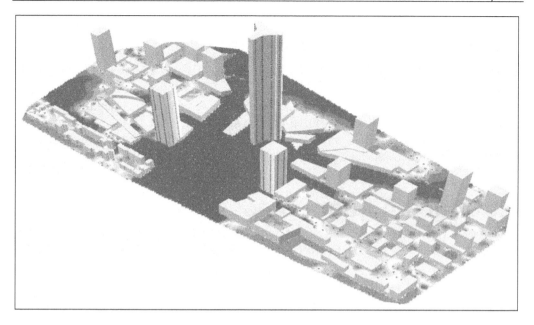

This book will not go into the details of the spatial analysis that you can perform with ArcGIS, but what is important to understand is that the ArcGIS platform has the capability to help solve problems such as these:

- An epidemiologist collects data on a disease, such as **Chronic Obstructive Pulmonary Disease (COPD)**, and wants to know where it occurs and whether there are any statistically significant clusters so that a mitigation plan can be developed

- A mining geologist wants to obtain samples of a precious mineral so that he/she can estimate the overall concentration of the mineral

- A military analyst or soldier wants to know where they can be located in the battlefield and not been seen

- A crime analyst wants to know where crimes are concentrated so that they can increase police presence as a deterrent

- A research scientist wants to develop a model to predict the path of a fire

There are many more examples. With ArcGIS Desktop and Pro, along with the correct extension, questions can be posed and answered using a variety of techniques. However, it's important to understand that ArcGIS Runtime may or may not be a good fit and may or may not support certain tools.

In many cases, spatial analysis would be best studied with ArcGIS Desktop or Pro. For example, if you plan to conduct hotspot analysis on patients or crime, doing this kind of operation with Desktop or Pro is best suited because it's typically something you do once. On the other hand, if you plan to allow users to repeat this process again and again with different data, and you need high performance, building a tool with ArcGIS Runtime will be the perfect solution, especially if they need to run the tool in the field. It should also be noted that, in some cases, the ArcGIS JavaScript API will also be better suited. It just goes back to the discussion we had in *Chapter 1, Introduction to ArcGIS Runtime*.

Introducing geoprocessing

If you open up the **Geoprocessing** toolbox in ArcGIS Desktop or Pro, you will find dozens of tools categorized in the following manner:

With these tools, you can build sophisticated models by using `ModelBuilder` or Python, and then publish them to ArcGIS Server. For example, to perform a buffer, as we did in *Chapter 5, Geometry and Symbology*, with the `GeometryEngine` tool, you would drag the `Buffer` tool onto the `ModelBuilder` canvas, as shown here, and specify its inputs and outputs:

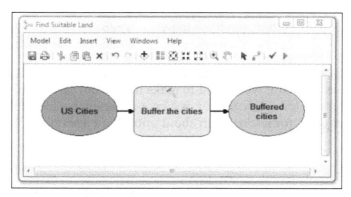

Drag tools on the canvas

This model specifies an input (**US cities**), performs an operation (**Buffer the cities**), and then produces an output (**Buffered cities**). Conceptually, this is programming, except that the algorithm is built graphically instead of with code. You may be asking: Why would you use this tool in ArcGIS Desktop or Pro? Good question. Well, ArcGIS Runtime only comes with a few selected tools in `GeometryEngine`. These tools, such as the buffer method in `GeometryEngine`, are so common that Esri decided to include them with ArcGIS Runtime so that these kinds of operation could be performed on the client without having to call the server. On the other hand, in order to keep the core of ArcGIS Runtime lightweight, Esri wanted to provide these tools and many more, but make them available as tools that you need to call on when required for special or advanced analysis. As a result, if your app needs basic operations, `GeometryEngine` may provide what you need. On the other hand, if you need to perform more sophisticated operations, you will need to build the model with Desktop or Pro, publish it to Server, and then consume the resulting service with ArcGIS Runtime. The rest of this chapter will show you how to consume a geoprocessing model using this pattern.

Preparing for geoprocessing

To perform geoprocessing, you will need to create a model with `ModelBuilder` and/or Python. For more details on how to create models using `ModelBuilder`, navigate to `http://pro.arcgis.com/en/pro-app/help/analysis/` `geoprocessing/modelbuilder/what-is-modelbuilder-.htm`.

To build a model with Python, navigate to `http://pro.arcgis.com/en/pro-app/` `help/analysis/geoprocessing/basics/python-and-geoprocessing.htm`.

Once you've created a model with `ModelBuilder` or Python, you will then need to run the tool to ensure that it works and to make it so that it can be published as a geoprocessing service for online use, or as a geoprocessing package for offline use. Use the following URL for information about publishing a service:

`http://server.arcgis.com/en/server/latest/publish-services/windows/a-` `quick-tour-of-publishing-a-geoprocessing-service.htm`

If you plan to use geoprocessing offline, you'll need to publish a geoprocessing package (`*.gpk`) file. You can learn more about these at `https://desktop.arcgis.` `com/en/desktop/latest/analyze/sharing-workflows/a-quick-tour-of-` `geoprocessing-packages.htm`.

Once you have a geoprocessing service or package, you can now consume it with ArcGIS Runtime. In the sections that follow, we will use classes from `Esri.ArcGISRuntime.Tasks.Geoprocessing` that allow us to consume these geoprocessing services or packages.

Online geoprocessing with ArcGIS Runtime

Once you have created a geoprocessing model, you will want to access it from ArcGIS Runtime. In this section, we're going to do surface analysis from an online service that Esri has published. To accomplish this, you will need to access the REST endpoint by typing in the following URL:

`http://sampleserver6.arcgisonline.com/arcgis/rest/services/Elevation/` `ESRI_Elevation_World/GPServer`

When you open this page, you'll notice the description and that it has a list of **Tasks**:

A task is a REST child resource of a geoprocessing service. A geoprocessing service can have one or more tasks associated with it. A task requires a set of inputs in the form of parameters. Once the task completes, it will produce some output that you will then use in your app. The output could be a map service, a single value, or even a report. This particular service only has one task associated with it and it is called Viewshed. If you click on the task called Viewshed, you'll be taken to this page: http://sampleserver6.arcgisonline.com/arcgis/rest/services/Elevation/ESRI_Elevation_World/GPServer/Viewshed.

This service will produce a viewshed of where the user clicks that looks something like this:

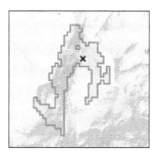

Viewshed of where the user clicks

The user clicks on the map (**X**) and the geoprocessing task produces a viewshed, which shows all the areas on the surface that are visible to an observer, as if they were standing on the surface.

Once you click on the task, you'll note the concepts marked in the following screenshot:

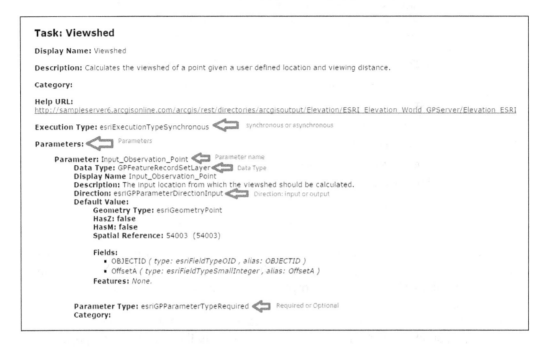

As you can see, beside the red arrows, the geoprocessing service lets you know what is required for it to operate, so let's go over each of these:

1. First, the service lets you know that it is a synchronous geoprocessing service. A synchronous geoprocessing task will run synchronously until it has completed, and block the calling thread. An asynchronous geoprocessing task will run asynchronously, but it won't block the calling thread.

2. The next pieces of information you'll need to provide to the task are the parameters. In the preceding example, the task requires `Input_Observation_Point`. You will need to provide this exact name when providing the parameter later on, when we write the code to pass in this parameter.

3. Also, note that the **Direction** value is `esriGPParameterDirectionInput`. This tells you that the task expects that `Input_Observation_Point` is an input to the model.

4. Lastly, note that the **Parameter Type** value is **Required**. In other words, you must provide the task with this parameter in order for it to run.

It's also worth noting that **Default Value** is an esriGeometryPoint type, which in ArcGIS Runtime is MapPoint. The **Spatial Reference** value of the point is **540003**. If you investigate the remaining required parameters, you'll note that they require a Viewshed_Distance parameter. Now, refer to the following screenshot. If you don't specify a value, it will use **Default Value** of 15,000 meters. Lastly, this task will output a Viewshed_Result parameter, which is esriGeometryPolygon. Using this polygon, we can then render to the map or scene.

```
Parameter: Viewshed_Distance    <=
    Data Type: GPLinearUnit
    Display Name Viewshed_Distance
    Description: The maximum distance from the input point for which the viewshed should be calculated. The maximum
    allowed distance is 20000 meters.
    Direction: esriGPParameterDirectionInput
    Default Value: 15000.0   (esriMeters)
    Parameter Type: esriGPParameterTypeRequired
    Category:

Parameter: Viewshed_Result    <=
    Data Type: GPFeatureRecordSetLayer
    Display Name Viewshed_Result
    Description: The resulting viewshed feature class given the user location and maximum distance.
    Direction: esriGPParameterDirectionOutput
    Default Value:
        Geometry Type: esriGeometryPolygon
        HasZ: false
        HasM: false
        Spatial Reference: 54003  (54003)

        Fields:
            • OBJECTID ( type: esriFieldTypeOID , alias: OBJECTID )
            • Id ( type: esriFieldTypeInteger , alias: Id )
            • grid_code ( type: esriFieldTypeInteger , alias: grid_code )
            • Shape_Length ( type: esriFieldTypeDouble , alias: Shape_Length )
            • Shape_Area ( type: esriFieldTypeDouble , alias: Shape_Area )
        Features: None.

    Parameter Type: esriGPParameterTypeRequired
    Category:
```

Geoprocessing synchronously

Now that you've seen an online service, let's look at how we call this service using ArcGIS Runtime. To execute the preceding viewshed task, we first need to create an instance of the Geoprocessor object. The Geoprocessor object requires a URL down to the task level in the REST endpoint, like this:

```
private const string viewshedServiceUrl =
    "http://sampleserver6.arcgisonline.com/arcgis/rest/services/
        Elevation/ESRI_Elevation_World/GPServer/Viewshed";
private Geoprocessor gpTask;
```

Note that we've attached /Viewshed on the end of the original URL so that we can pass in the completed path to the task. Next, you will instantiate the geoprocessor in your app, using the URL to the task:

```
gpTask = new Geoprocessor(new Uri(viewshedServiceUrl));
```

Once we have created the geoprocessor, we can then prompt the user to click somewhere on the map. Let's look at some code:

```
public async void CreateViewshed()
{
    // // get a point from the user
    var mapPoint = await this.mapView.Editor.RequestPointAsync();

    // clear the graphics layers
    this.viewshedGraphicsLayer.Graphics.Clear();
    this.inputGraphicsLayer.Graphics.Clear();

    // add new graphic to layer
    this.inputGraphicsLayer.Graphics.Add(new Graphic{ Geometry =
        mapPoint, Symbol = this.sms });

    // specify the input parameters
    var parameter = new GPInputParameter() { OutSpatialReference =
        SpatialReferences.WebMercator };
    parameter.GPParameters.Add(new
        GPFeatureRecordSetLayer("Input_Observation_Point",
            mapPoint));
    parameter.GPParameters.Add(new
        GPLinearUnit("Viewshed_Distance",
            LinearUnits.Miles, this.distance));

    // Send to the server
    this.Status = "Processing on server...";
    var result = await gpTask.ExecuteAsync(parameter);
    if (result == null || result.OutParameters == null ||
        !(result.OutParameters[0] is GPFeatureRecordSetLayer))
        throw new ApplicationException("No viewshed graphics
            returned
                for this start point.");

    // process the output
    this.Status = "Finished processing. Retrieving results...";
    var viewshedLayer = result.OutParameters[0] as
        GPFeatureRecordSetLayer;
    var features = viewshedLayer.FeatureSet.Features;
    foreach (Feature feature in features)
```

```
    {
        this.viewshedGraphicsLayer.Graphics.Add(feature as
            Graphic);
    }
    this.Status = "Finished!!";
}
```

The first thing we do is have the user click on the map and return `MapPoint`. We then clear a couple of `GraphicsLayers` that hold the input graphic and viewshed graphics, so that the map is cleared every time they run this code. Next, we create a graphic using the location where the user clicked. Now comes the interesting part of this. We need to provide the input parameters for the task and we do that with `GPInputParameter`. When we instantiate `GPInputParameter`, we also need to specify the output spatial reference so that the data is rendered in the spatial reference of the map. In this example, we're using the map's spatial reference.

Then, we add the input parameters. Note that we've spelled them exactly as the task required them. If we don't, the task won't work. We also learned earlier that this task requires a distance, so we use `GPLinearUnit` in `Miles`. The `GPLinearUnit` class lets the geoprocessor know what kinds of unit to accept.

After the input parameters are set up, we then call `ExecuteAsync`. We are calling this method because this is a synchronous geoprocessing task. Even though this method has `Async` on the end of it, this applies to .NET, not ArcGIS Server. The alternative to `ExecuteAsync` is `SubmitJob`, which we will discuss shortly. After some time, the result comes back and we grab the results using `result.OutParameters[0]`. This contains the output from the geoprocessing task and we want to use that to then render the output to the map. Thankfully, it returns a read-only set of polygons, which we can then add to `GraphicsLayer`.

If you don't know which parameter to use, you'll need to look it up on the task's page. In the preceding example, the parameter was called `Viewshed_Distance` and the **Data Type** value was **GPLinearUnit**. ArcGIS Runtime comes with a variety of data types to match the corresponding data type on the server. The other supported types are `GPBoolean`, `GPDataFile`, `GPDate`, `GPDouble`, `GPItemID`, `GPLinearUnit`, `GPLong`, `GPMultiValue<T>`, `GPRasterData`, `GPRecordSet`, and `GPString`.

Instead of manually inspecting a task, as we did earlier, you can also use `Geoprocessor.GetTaskInfoAsync` to discover all of the parameters. This is a useful object if you want to provide your users with the ability to specify any geoprocessing task dynamically while the app is running. For example, if your app requires that users are able to enter any geoprocessing task, you'll need to inspect that task, obtain the parameters, and then respond dynamically to the entered geoprocessing task.

If you open up the sample code that came with this book, you will find a project called Chapter10, which shows the app running the Viewshed task. The app follows the MVVM pattern, just like all the other apps that came with this book. Feel free to change the color and symbols for the input point and viewshed polygon.

Geoprocessing asynchronously

So far we've called a geoprocessing task synchronously. In this section, we'll cover how to call a geoprocessing task asynchronously. There are two differences when calling a geoprocessing task asynchronously:

- You will run the task by executing a method called SubmitJobAsync instead of ExecuteAsync. The SubmitJobAsync method is ideal for long-running tasks, such as performing data processing on the server. The major advantage of SubmitJobAsync is that users can continue working while the task works in the background. When the task is completed, the results will be presented.

- You will need to check the status of the task with GPJobStatus so that users can get a sense of whether the task is working as expected. To do this, check GPJobStatus periodically and it will return GPJobStatus. The GPJobStatus enumeration has the following values: New, Submitted, Waiting, Executing, Succeeded, Failed, TimedOut, Cancelling, Cancelled, Deleting, or Deleted. With these enumerations, you can poll the server and return the status using CheckJobStatusAsync on the task and present that to the user while they wait for the geoprocessor.

Let's take a look at this process in the following diagram:

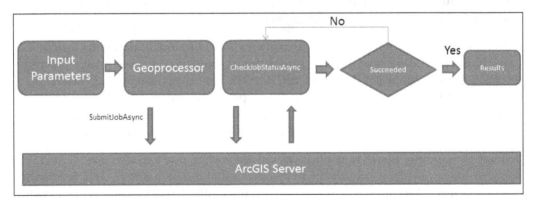

As you can see in the preceding diagram, the input parameters are specified as we did earlier with the synchronous task, the **Geoprocessor** object is set up, and then `SubmitJobAsync` is called with the parameters (`GpInputParameter`). Once the task begins, we then have to check the status of it using the results from `SubmitJobAsync`. We then use `CheckJobStatusAsync` on the task to return the status enumeration. If it indicates Succeeded, we then do something with the results. If not, we continue to check the status using any time period we specify.

Let's try this out using an example service from Esri that allows for areal analysis. Go to the following REST endpoint: `http://serverapps10.esri.com/ArcGIS/rest/services/SamplesNET/USA_Data_ClipTools/GPServer/ClipCounties`.

In the service, you will note that it's called `ClipCounties`. This is a rather contrived example, but it shows how to do server-side data processing. It requires two parameters called `Input_Features` and `Linear_unit`. It outputs `output_zip` and `Clipped _Counties`. Basically, this task allows you to drag a line on the map; it will then buffer it and clip out the counties in the U.S. and show them on the map, like so:

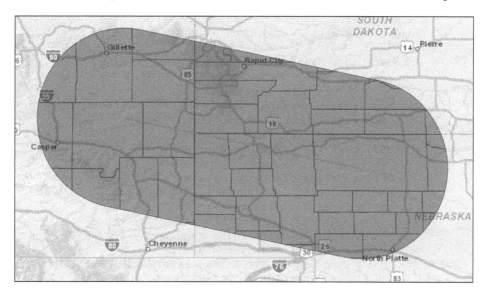

If you take a look at the sample apps that come with this book, you'll find an app called `Chapter10a`. We are interested in two methods in this sample app. Let's take a look at them:

```
public async void Clip()
{
    //get the user's input line
    var inputLine = await this.mapView.Editor.RequestShapeAsync(
        DrawShape.Polyline) as Polyline;
```

```
// clear the graphics layers
this.resultGraphicsLayer.Graphics.Clear();
this.inputGraphicsLayer.Graphics.Clear();

// add new graphic to layer
this.inputGraphicsLayer.Graphics.Add(
    new Graphic { Geometry = inputLine, Symbol =
        this.simpleInputLineSymbol });

// add the parameters
var parameter = new GPInputParameter();
parameter.GPParameters.Add(
    new GPFeatureRecordSetLayer("Input_Features", inputLine));
parameter.GPParameters.Add(new GPLinearUnit(
    "Linear_unit", LinearUnits.Miles, this.Distance));

// poll the task
var result = await SubmitAndPollStatusAsync(parameter);

// add successful results to the map
if (result.JobStatus == GPJobStatus.Succeeded)
{
    this.Status = "Finished processing. Retrieving
        results...";

    var resultData = await
        gpTask.GetResultDataAsync(result.JobID,
            "Clipped_Counties");
    if (resultData is GPFeatureRecordSetLayer)
    {
        GPFeatureRecordSetLayer gpLayer =
            resultData as GPFeatureRecordSetLayer;
        if (gpLayer.FeatureSet.Features.Count == 0)
        {
            // the the map service results
            var resultImageLayer = await
                gpTask.GetResultImageLayerAsync(
                    result.JobID, "Clipped_Counties");

            // make the result image layer opaque
            GPResultImageLayer gpImageLayer =
                resultImageLayer;
            gpImageLayer.Opacity = 0.5;
            this.mapView.Map.Layers.Add(gpImageLayer);
            this.Status = "Greater than 500 features returned.
                Results drawn using map service.";
            return;
        }
```

```
        // get the result features and add them to the
        // GraphicsLayer
        var features = gpLayer.FeatureSet.Features;
        foreach (Feature feature in features)
        {
            this.resultGraphicsLayer.Graphics.Add(
                feature as Graphic);
        }

    }
    this.Status = "Success!!!";
  }
}
```

This `Clip` method first asks the user to add a polyline to the map. It then clears the `GraphicsLayer` class, adds the input line to the map in red, sets up `GPInputParameter` with the required parameters (`Input_Featurs` and `Linear_unit`), and calls a method named `SubmitAndPollStatusAsync` using the input parameters. Let's take a look at that method too:

```
// Submit GP Job and Poll the server for results every 2 seconds.
private async Task<GPJobInfo>
    SubmitAndPollStatusAsync(GPInputParameter parameter)
{
    // Submit gp service job
    var result = await gpTask.SubmitJobAsync(parameter);

    // Poll for the results async
    while (result.JobStatus != GPJobStatus.Cancelled &&
        result.JobStatus != GPJobStatus.Deleted     &&
        result.JobStatus != GPJobStatus.Succeeded &&
        result.JobStatus != GPJobStatus.TimedOut)
    {
        result = await gpTask.CheckJobStatusAsync(result.JobID);

        foreach (GPMessage msg in result.Messages)
        {
            this.Status = string.Join(Environment.NewLine,
                msg.Description);
        }
        await Task.Delay(2000);
    }

    return result;
}
```

The `SubmitAndPollStatusAsync` method submits the geoprocessing task and the polls it every two seconds to see if it hasn't been `Cancelled`, `Deleted`, `Succeeded`, or `TimedOut`. It calls `CheckJobStatusAsync`, gets the messages of type `GPMessage`, and adds them to the property called `Status`, which is a ViewModel property with the current status of the task. We then effectively check the status of the task every 2 seconds with `Task.Delay(2000)` and continue doing this until something happens other than the `GPJobStatus` enumerations we're checking for.

Once `SubmitAndPollStatusAsync` has succeeded, we then return to the main method (`Clip`) and perform the following steps with the results:

1. We obtain the results with `GetResultDataAsync` by passing in the results of `JobID` and `Clipped_Counties`. The `Clipped_Counties` instance is an output of the task, so we just need to specify the name `Clipped_Counties`.

2. Using the resulting data, we first check whether it is a `GPFeatureRecordSetLayer` type. If it is, we then do some more processing on the results.

3. We then do a cast just to make sure we have the right object (`GPFeatureRecordsSetLayer`).

4. We then check to see if no features were returned from the task. If none were returned, we perform the following steps:

    1. We obtain the resulting image layer using `GetResultImageLayerAsync`. This returns a map service image of the results.

    2. We then cast this to `GPResultImageLayer` and set its opacity to `0.5` so that we can see through it. If the user enters in a large distance, a lot of counties are returned, so we convert the layer to a map image, and then show them the entire country so that they can see what they've done wrong. Having the result as an image is faster than displaying all of the polygons as the JSON objects.

    3. Add `GPResultImageLayer` to the map.

5. If everything went according to plan, we get only the features needed and add them to `GraphicsLayer`.

That was a lot of work, but it's pretty awesome that we sent this off to ArcGIS Server and it did some heavy processing for us so that we could continue working with our map. The geoprocessing task took in a user-specified line, buffered it, and then clipped out the counties in the U.S. that intersected with that buffer. When you run the `Chapter10a` project, make sure you pan or zoom around while the task is running so that you can see that you can still work. You could also further enhance this code to zoom to the results when it finishes.

There are some other pretty interesting capabilities that we need to discuss with this code, so let's delve a little deeper.

Working with the output results

Let's discuss the output of the geoprocessing results in a little more detail in this section.

GPMesssage

The GPMessage object is very helpful because it can be used to check the types of message that are coming back from the Server. It contains different kinds of message via an enumeration called GPMessageType, which you can use to further process the message. GPMessageType returns an enumeration of Informative, Warning, Error, Abort, and Empty. For example, if the task failed, GPMessageType.Error will be returned and you can present a message to the user letting them know what happened and what they can do to resolve this issue. The GPMessage object also returns Description, which we used in the preceding code to display to the user as the task executed. The level of messages returned by the Server dictates what messages are returned by the task. Look at **Message Level** in the following screenshot:

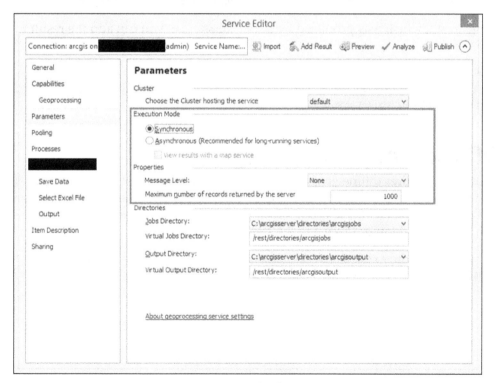

Message level

If the **Message Level** field is set to **None**, no messages will be returned. When testing a geoprocessing service, it can be helpful to set the service to **Info** because it produces detailed messages.

GPFeatureRecordSetLayer

The preceding task expected an output of features, so we cast the result to GPFeatureRecordsSetLayer. The GPFeatureRecordsSetLayer object is a layer type which handles the JSON objects returned by the server, which we can then use to render on the map.

GPResultMapServiceLayer

When a geoprocessing service is created, you have the option of making it produce an output map service result with its own symbology. Refer to http://server. arcgis.com/en/server/latest/publish-services/windows/defining-output-symbology-for-geoprocessing-tasks.htm.

You can take the results of a GPFeatureRecordsSetLayer object and access this map service using the following URL format:

http://catalog-url/resultMapServiceName/MapServer/jobs/jobid

Using JobID, which was produced by SubmitJobAsync, you can add the result to the map like so:

```
ArcGISDynamicMapServiceLayer dynLayer =
    this.gpTask.GetResultMapServiceLayer(result.JobID);
this.mapView.Map.Layers.Add(dynLayer);
```

Offline geoprocessing

If your app requires offline geoprocessing, you can use the LocalGeoprocessingService instance, by using the path to the local geoprocessing package. To create an offline geoprocessing package, you first need to create one with ArcGIS Desktop or Pro using the steps outlined here:

https://desktop.arcgis.com/en/desktop/latest/analyze/sharing-workflows/a-quick-tour-of-creating-a-geoprocessing-package.htm

The first thing to note is that `LocalGeoprocessingService` is located in `Esri.ArcGISRuntime.LocalServices`. Running an offline geoprocessing service is very similar to how you would use an online geoprocessing service. However, instead of just accessing the service using an online URL, you will need to start up the local geoprocessing task on your machine, like this:

```
this.localGPService = new
    LocalGeoprocessingService(this.clipGPKPath,
    GeoprocessingServiceType.SubmitJob);
this.localGPService.StartAsync();
```

Note that we've instantiated `LocalGeoprocessingService` with the path to the geoprocessing package (`*.gpk`) and we tell it that this is an asynchronous task because we're using `SubmitJob`. Then, we start the service. This is a very similar in concept to the `LocalMapService` class we created in *Chapter 3, Maps and Layers*. In effect, `LocalGeoprocessingService` and `LocalMapService` provide us with a local ArcGIS Server on a PC. Note that `LocalGeoprocessingService` is not available in the Windows Store or Windows Phone versions of the ArcGIS Runtime SDK for .NET.

When we start this service, it will create a REST endpoint just like the ones you've been using with online tasks. The big difference is that the REST endpoint isn't generated until you start the task with `StartAsync`. If you look at the sample code that came with this book, you'll note in the sample chapter called `Chapter10b` that the instantiation of the geoprocessor has been moved to the `Clip` method, so that when the user clicks on the **Start** button, the geoprocessor is instantiated and passed in the URL of the task, which comes from the `UrlGeoprocessingService` property of `LocalGeoprocessingService`:

```
public async void Clip()
{
    this.gpUrl = this.localGPService.UrlGeoprocessingService;

    this.gpTask = new Geoprocessor(new Uri(this.gpUrl +
        "/ClipFeatures"));

    // removed for brevity
}
```

If you place a break point on the line that starts with `this.gpTask`, you will note that the line before it generated a URL for you and it will be similar to this URL: `http://127.0.0.1:50000/k7sT0w/arcgis/rest/services/clip-features/GPServer/ClipFeatures` (yours will be different).

While the app is running, you can open your favorite browser and inspect this URL (without /ClipFeatures in the end). You'll find a page like this:

Notice anything different? Well, there are a few differences. First, with this task, Input_Features is called **Input**. Also, Linear_unit is now called **Linear_Unit**. Other than those minor differences, the service is the same as the online version of the task. Also, as soon as you close the app, this service will no longer be available because LocalGeoprocessingService will be stopped.

Execution types

The local geoprocessing server has the same kinds of execution mode as ArcGIS Server. However, you set when to instantiate the LocalGeoprocessingService instance using GeomprocessingServiceType.Execute, GeomprocessingServiceType.SubmitJob, or GeomprocessingServiceType. SubmitJobMapServerResult. The first two instances have been discussed. The SubmitJobMapServerResult instance is the same as SubmitJob, except that an additional local map service is instantiated. The SubmitJobMapServerResult instance is useful if LocalGeoprocessingService is returning a large number of features, which allows you to control the symbology however you want. This also improves performance.

Maximum number of records

When you use an online or offline geoprocessing task, it will by default only return 1,000 features. The reason for this is to improve performance. With most apps, this setting is a good number to use. However, if your app will return more than this, you can have your ArcGIS Server administrator change this setting when the service is created or modified after it has been created, by changing the setting in the service called `Maximum Number of Records Returned by Server`. See the following website for how to do this:

```
http://server.arcgis.com/en/server/latest/publish-services/windows/
geoprocessing-service-settings-parameters.htm
```

With the `LocalGeoprocessingService` instance, you'll need to set a property called `MaxRecords`.

Geoprocessing tools available to ArcGIS Runtime

Now that you've learned how to do geoprocessing with ArcGIS Runtime, we need to cover what you can and can't use from ArcGIS Desktop/Pro. As we discussed earlier in this book, ArcGIS Runtime is not a desktop tool; it's for building tightly focused apps. As such, only a subset of geoprocessing tools is available for use in your apps. For a complete list, see here:

```
https://developers.arcgis.com/net/desktop/guide/local-server-
geoprocessing-tools-support.htm
```

Summary

In this chapter, we went over spatial analysis at a high level, and then went into the details of how to do spatial analysis with ArcGIS Runtime. We discussed how to create models with `ModelBuilder` and/or Python, and then went on to show how to use geoprocessing, both synchronously and asynchronously, with online and offline tasks. With this information, you now have a multitude of options for adding a wide variety of analytical tools to your apps.

Now that we've covered the most important features of the API, we will turn our attention to testing and performance considerations in the next chapter.

11

Testing and Performance

Now that we've surveyed the API and SDK of ArcGIS Runtime, it's time to turn our attention to testing our app, while at the same time making sure that it is performant. Test and performance tend to go together because as you test your app you'll discover performance problems. In this chapter, we'll discuss testing and performance so that you will gain a solid understand of the concepts and techniques required to address these two important subjects. We will first learn about unit testing in general, and then learn how to do unit testing with MVVM via MVVM Light. We will also learn how to make sure our app runs as quickly as possible, both from the server side and the client side. In summary, the following topics will be addressed:

- General unit testing
- Using testing with ArcGIS Runtime
- Using testing with MVVM Light
- Test-driven development
- Unit testing with Windows Store
- Service-side performance
- Client-side performance

Testing

Testing your apps is critical to the success of your apps' acceptance by users. Even if 99 percent of your app works perfectly, it only takes one error to kill your apps' success. Not only does it leave a bad taste in your users' mouths, it can also cost you a lot of money to fix the bugs after the app has been released. One very beneficial strategy for resolving errors before they arise is to test your app using a variety of techniques.

There are many testing techniques you will need to employ in order to make your app as bulletproof as possible:

- **Installation testing**: This testing ensures that the app installs on customers' hardware. As ArcGIS Runtime supports many platforms, this could involve testing on the Windows devices, iOS devices, Android devices, and so on.

- **Compatibility testing**: This testing ensures that the app is compatible with the OS and other software.

- **Regression testing**: This testing is done after a major change in the code.

- **Acceptance testing**: This testing involves whether the user accepts that the app has been developed as agreed.

- **Software performance testing**: This testing ensures that the app performs as expected.

- **Development testing**: This testing is done by the development team, which involves having the developer test their classes at the atomic level (methods and properties).

- **Security testing**: This testing involves making sure that the app protects sensitive data and prevents hackers from intruding in the system.

This book was written for developers, so we are going to focus on the most fundamental unit of testing. That is, we're going to focus on unit testing the methods and properties of our classes. Later on, we'll address what you need to know when it comes to performance, so you can test your apps to ensure that they run as quickly as possible.

Unit testing

Unit testing is a methodology of testing the individual methods and properties of your classes using automated tools built into Visual Studio. In order to unit test, you will need a testing harness, and thankfully, Visual Studio provides a means to test your classes' methods and properties in an automated fashion.

See the following image:

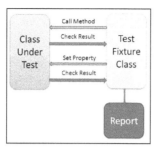

In this diagram, there is a **Test fixture Class** block that calls on your class to test it. The test fixture class is first instantiated with whatever it needs, you then call the method or property, and then then the results are returned. You use the returned results to ensure that the method or property worked as expected. Visual Studio will then present the results of the test in the form of a simple report.

Unit testing ArcGIS Runtime

Now that we conceptually understand what unit testing is, let's try this out with a simple app in Visual Studio:

1. Open Visual Studio and create a new project. Click on **Test** and then **Unit Test Project**, as shown here:

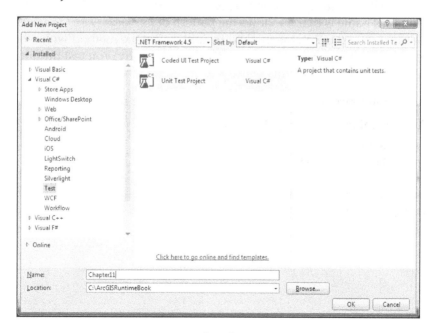

2. Give the project a name and click on **OK**. The following class will be created:

```
using System;
using Microsoft.VisualStudio.TestTools.UnitTesting;

namespace Chapter11
{
    [TestClass]
    public class UnitTest1
    {
        [TestMethod]
        public void TestMethod1()
        {
        }
    }
}
```

This is the test fixture class. We will now use this class to test some generic ArcGIS Runtime code so that we can learn about unit testing.

3. Rename the method called `TestMethod1`, and name it `GetRedlineCountAsync_CountFeatures_FiveFeaturesFound`.

That's a rather long name, but there's a good reason for naming it that. When conducting unit testing, the basic idea is that eventually you will have lots of methods in your app and you want to give them unique names, because as the number of unit tests grows, you'll need unique names to help you find the correct test. Another thing to note about this name is that it follows the pattern of `Arrange`/`Act`/`Assert`. The basic idea is that you `Arrange` the method by initializing the method with whatever it needs to execute. Then, you execute the method (`Act`). Lastly, you test the outcome of the unit test (`Assert`). Let's put this to use with a simple test:

1. Add a reference to `Esri.ArcGISRuntime` and `WindowsBase`.

2. Add the following `using` statements at the top of your test class:

```
using System.Threading.Tasks;
using Esri.ArcGISRuntime.Controls;
using Esri.ArcGISRuntime.Data;
using Esri.ArcGISRuntime.Layers;
using Esri.ArcGISRuntime.Tasks.Query;
```

3. In the new method named GetRedlineCountAsync_CountFeatures_
   FiveFeaturesFound, add the following code:

```
[TestClass]
public async Task
    GeteRedlineCountAsync_CountFeatures_FiveFeaturesFound()
{
    QueryCountResult results = null;
    Map map = null;

    var table = new ServiceFeatureTable(
        new
            Uri("http://sampleserver6.arcgisonline.com/
                arcgis/rest/services/Water_Network/
                    FeatureServer/2"));

    await table.InitializeAsync();
    map = new Map();
    map.Layers.Add(new FeatureLayer(table) { ID = "test"
        });


    var queryTask = new QueryTask(new
        Uri(table.ServiceUri));
    results = await queryTask.ExecuteCountAsync(new
        Query("1=1"));

    Assert.AreEqual(5, results.Count);
}
```

4. From the main menu in Visual Studio, navigate to **Test | Windows | Test Explorer**. A window will appear. See here:

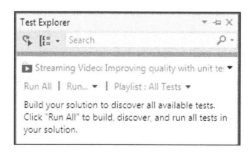

5. Click on **Run All** in the **Test Explorer** window. If everything went OK, you will see results similar to the following screenshot:

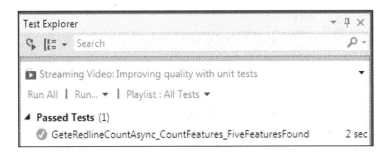

Although this was a rather contrived example, it illustrates unit testing. Let's review exactly how this worked. The first thing you probably noted is that there are the .NET attributes added before the class name and method: [TestClass] and [TestMethod]. If you look at this project's type, you'll note that it is a class library, but by using these attributes, we are telling Visual Studio that this is a test library, which allows us to run this unit test. Other than that minor difference, it's just like any other class library. We just need to assign the [TestClass] attribute in front the class name, and [TestMethod] before any method we want to use.

The next thing to note is how we set up this test method. We first created QueryCountResult to hold the results of ExecuteCountAsync. We then created a Map variable, opened ServiceFeatureTable to an online service, initialized it, created a map instance, added the layer to the map, and then queried it using an expression ("1=1") that would return all the features in the layer. In unit testing parlance, we arranged, acted, and asserted all with a single test method. We could have arranged the test outside the test method, but for now we did everything within the test method.

The Assert.AreEqual call is the part that we really wanted to test. It is a simple static class that allows us to assert that something is true or false. In this case, we are asserting that this feature service has five records in it. We may want to actually test for this if our app depends on data residing in this online service. Furthermore, there are other kinds of assertion that you can make with Visual Studio, such as Assert.IsTrue, Assert.AreSame, and Assert.IsNotNull. You can use all of these assertions to fully exercise your methods so that you can ensure they work properly before you use them in your app.

However, it's recommended that you use one `Assert` instance per method, so that your tests only do one thing. This will aid you in finding bugs, especially as the number of tests grows.

Lastly, when we ran the test from the **Test Explorer** window, it lets us know that the test succeeded with a green check mark beside the unit test name, and it lets us know this under the heading `Passed Tests`.

The example app for this section of the book is available with the sample code that came with this book; it's called `Chapter11`.

With this information, we can now build more complex tests, such as testing that a method on our ViewModel works correctly.

Unit testing with MVVM Light

Unit testing with MVVM Light takes a little more work. In this section, we're going to discuss what we need to do first, and then implement a simple solution for the project we built in *Chapter 4, From 2D to 3D*. In that chapter, we built a 3D app that allowed the user to search for a city, county, or state using a search string. Don't worry about the 3D part, because this equally applies to a 2D app too. Do the following steps:

1. Create a unit test project, as you did in the previous section. In this section, the project will be named `Chaper11a`.

2. Add references to the following `Esri.ArcGISRuntime` namespace and MVVM Light libraries only, `PresentationCore`, `PresentationFramework`, `System.Xaml`, and `WindowsBase`:

```
using System;
using System.Threading.Tasks;
using System.Windows;
using Microsoft.VisualStudio.TestTools.UnitTesting;
using System.Collections.ObjectModel;
using System.Threading;

using Esri.ArcGISRuntime.Controls;
using Esri.ArcGISRuntime.Data;
using Esri.ArcGISRuntime.Layers;
using Esri.ArcGISRuntime.Tasks.Query;

using GalaSoft.MvvmLight.Messaging;

using Chapter4a.ViewModels;
```

Note that we added the required references, and a reference to the `Chapter4a` project. This is accomplished by adding a project reference like so:

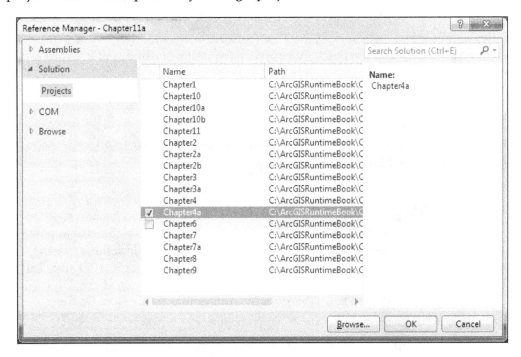

Note that you navigate to **Solution** | **Projects**, and then click on `Chapter4a`.

One of the limitations of MVVM Light is that it doesn't allow you to run a `RelayCommand` method asynchronously. This is especially problematic for unit testing because you want to be able to automate your tests. We could run them and wait for them to complete using `Thread.Sleep` or `Task.Delay`, but neither of those options are very good solutions. They simply make your unit tests run far longer than they should. Also, we could refactor the source code to run synchronously, but we want it to run asynchronously as it was written. What we want to do is run the `Search` method in this project asynchronously. Unfortunately, MVVM Light's `RelayCommand` method doesn't allow this. The good news is that we can write our own asynchronous `RelayCommand` method and this will allow us to unit test this app with very little modification.

This book comes with an asynchronous `RelayCommand` method that we will now use. The next thing we need to do is refactor our code so that we can run this new asynchronous `RelayCommand` method. In the project named `Project4a`, you'll find a file called `RelayCommandAsync.cs`:

1. In `MainViewModel.cs`, add a `using` statement to `MvvmHelper`.

2. Change the type of `SearchRelayCommand` to `RelayCommandAsync`.

3. In the constructor, change this line:

   ```
   this.SearchRelayCommand = new RelayCommand<int>(Search);
   ```

4. Change it to this:

   ```
   this.SearchRelayCommand = new
       RelayCommandAsync<int>(Search);
   ```

5. Add the using statement of `System.Threading.Tasks` to `MainViewModel.cs`.

6. Change the return type of the **Search** method from `void` to the following code:

   ```
   public async Task Search(int wkid)
   ```

   This is necessary in order for the method to run asynchronously. Also, using `Task` as the return type is Microsoft's recommended approach.

7. Rebuild your project if you're following this step by step.

8. Now that we have the references set up and an asynchronous `RelayCommand` method, let's add the unit testing code. All of the test class is included here so that you can see everything that needs to be entered:

   ```
   namespace Chapter11a
   {
       [TestClass]
       public class UnitTest1
       {
           private ViewModelLocator locator = null;

           [TestMethod]
           public   void
               public   async Task
               GetSearchCount_CountFeatures_NineFeatureFound()
   ```

```
        {
            // instantiate the locator so that our view
                model is
            // created
            locator = new ViewModelLocator();
            // get the MainViewModel
            MainViewModel mainViewModel =
                locator.MainViewModel;
            // create a MapView
            MapView mapView = new MapView();
            // send the MapView to the MainViewModel
            Messenger.Default.Send<MapView>(mapView);

            // Arrange
            // search string
            mainViewModel.SearchText = "Lancaster";
            // search using the search text and SR ID
            // run the search async
            var task =
                mainViewModel.SearchRelayCommand.
                    ExecuteAsync(4326);
                task.Wait(); // wait

                // Assert
                Assert.IsNotNull(mainViewModel, "Null
                    mainViewModel");
                Assert.IsNotNull
                    (mainViewModel.GridDataResults,
                    "Null GridDataResults");
                Assert.AreEqual(9,
                    mainViewModel.GridDataResults.Count);

        }
    }
}
```

9. From the **Text Explorer** window, right-click on the unit test called **GetSearchCount...** and click on **Run Selected Test**. After a brief period of time, you should see a message box appear. Click on **OK**. Then, you will see that the test passed.

The first thing we have going on is the test method first instantiates `ViewModelLocator`. In essence, this is what happens in all of the other apps we've created. The `ViewModelLocator` method is instantiated and that provides our app with a reference to `MainViewModel`. The next thing we do is create an instance of `MapView`. In the apps we've been using, we did this through a custom behavior, but in this example we're mocking this process in order to fool our `ViewModel` class into thinking it has `MapView`. Then, we send `MapView` via `Messenger` to the listener, which in this case is `MainViewModel`.

After we've set things up (`Arrange`), we now set up a property on the `ViewModel` class with the search string. We then call the asynchronous `SearchRelayCommand` method and pass in the spatial reference we used in that project. We then wait for it to complete with `Task.Wait()`. After it completes, we first assert that the `ViewModel` class isn't `null`, assert that the property called `GridDataResults` isn't `null`, and that the `Count` property on `GridDataResults` isn't `null`, and finally we assert that there are nine features returned by the `Search` method. Since this is the number of features in this layer, it returns `true` and our test passes.

With all that said, you now have a means to unit test a Windows app. What's more is that you've learned about an important limitation of MVVM Light, but you now know how to resolve this with future apps that you write for Windows. You also saw that we needed to refactor our code to handle the asynchronous `RelayCommand` methods and we changed our method to return `Task`, which is the recommended approach. For more details on Microsoft best practices with regards to asynchronous programming, see here:

https://msdn.microsoft.com/en-us/magazine/jj991977.aspx

Lastly, this simple app displays a message box when it is complete. In a professional app, we would not use this approach. We'd display the message in a message window or as a popup that only displays for a few seconds, similar to **Toast** in Android. Otherwise, the message box would prevent our unit test from being automated.

This brings up an important point. We should test our apps as part of the development so that these kinds of problem are addressed early in the design.

Test-driven development

The examples we've discussed here so far have been high-level. We should actually write some other unit tests that test for `null` values, invalid properties, and so on, so that we can find bugs in our code. But, more importantly, we really should test for badly-designed components. The **test-driven development** (TDD) process is where unit testing shines. TDD is a powerful way of designing software components in an interactive way, so that their behavior is specified through the unit test. TDD helps you write software components that individually behave as designed. With TDD, you can easily refactor your code and it provides a way to document your code as you write it. Ultimately, what's great about TDD is that our components have been designed from the ground up—at the unit level. This way, we can make changes to the component, and as long as we don't change the interface, all other components should have no idea that it has changed. For more information on TDD, see here:

```
https://msdn.microsoft.com/en-us/library/aa730844(v=vs.80).aspx
```

Code coverage

As part of your development efforts, you will want to make sure that you test as much of your code as possible. Your ultimate goal is to achieve 100 percent of the methods and properties of your classes, by checking that they execute as expected in your unit tests. While this is a laudable goal, it is sometimes not really feasible given the design of the app and schedule. Regardless, you will want to make sure you checked that every method has been called, every statement has been executed, every branch (the `if` and `case` statements) has been executed, and lastly every Boolean expression has been evaluated as `true` or `false`. For more information on code coverage with Visual Studio, see here:

```
https://msdn.microsoft.com/en-us/library/dd537628(v=vs.120).aspx
```

Unit testing with Windows Store

Now that you've seen some unit testing with the Windows version of ArcGIS Runtime, let's try doing the same thing with a Windows Store version of ArcGIS Runtime.

You will need Windows 8.1 or Windows 10 to complete this exercise:

1. In Visual Studio, create **Unit Test Library**, as shown here:

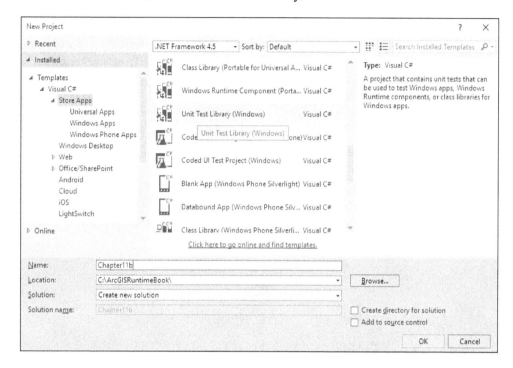

2. Add the same code as you did in the *Unit Test ArcGIS Runtime* of this chapter.

3. Run the app.

You will note that the following error occurs:

```
System.Exception: The application called an interface that was
    marshalled for a different thread. (Exception from HRESULT:
        0x8001010E (RPC_E_WRONG_THREAD))
```

This error occurs because we are trying to run ArcGIS Runtime in a class library, when in fact, it contains controls from `Esri.ArcGISRuntime.Controls`, which were meant to be run from the main UI thread.

Fortunately, there is a solution to this problem that was developed by Esri.

1. Open up the `Chapter11b` project that came with this book. In this project, you'll note a file called `Helpers\ThreadHelper.cs` and `SetupTestEnvironment.cs`. The `ThreadHelper` class allows you to run the preceding code on the main UI thread. In order for this class to work, it must first be initialized when the assembly loads. The `SetupTestEnvironment` instance has a `static` method that is called when the assembly is loaded. It gets the main view's dispatcher, and then initializes it. Once that is done, we can call our code using the `ThreadHelper` class. Copy `ThreadHelper.cs` and `SetupTestEnvironment.cs` to your project. Update namespaces if necessary.

2. Modify your code as shown here:

```
[TestClass]
public async Task
    GeteRedlineCountAsync_CountFeatures_FiveFeaturesFound()
{
    await ThreadHelper.Run(async () =>
    {

        QueryCountResult results = null;
        Map map = null;

        var table = new ServiceFeatureTable(
        new
            Uri("http://sampleserver6.arcgisonline.com/
                arcgis/rest/services/Water_Network/
                    FeatureServer/2"));

        await table.InitializeAsync();
        map = new Map();
        map.Layers.Add(new FeatureLayer(table) { ID =
            "test" });


        var queryTask = new QueryTask(new
            Uri(table.ServiceUri));
        results = await queryTask.ExecuteCountAsync(new
            Query("1=1"));

        Assert.AreEqual(5, results.Count);
    });
}
```

3. Run the app again and you'll note that the test succeeds without an error.

With this code, you can now unit test your apps that are targeted for Windows Store.

Performance

Now that we've learned about unit testing, let's discuss another important topic: performance. Your app needs to be responsive, or your users will simply not use it or will complain about it, which results in negative press for you or your company. The good news is that Esri has developed ArcGIS Runtime to be performant, both on the server side and the client side.

When speaking of the server side, we are referring to the fact that ArcGIS Runtime can use online content from a wide variety of sources. However, you as the developer can only control those web services that you have either developed yourself or use in your organization. As a result, we will limit our discussion to on-premises ArcGIS Server. We also won't discuss much about what we can do with ArcGIS Online sources such as basemaps, because we can assume that Esri scales it up as more users make requests for it. It is after all built on top of Amazon Web Services, which scales dynamically based on usage. We do, on the other hand, have some or complete control over an on-premises ArcGIS Server environment.

On the client side of things, we also have a great deal of control when working with the API. We can control what layers to display, how to display them, what fields to query, and so on. In this section, we're going to discuss most of the options that we have, in order to make our apps as responsive as possible while at the same time considering the costs associated with these choices. When we refer to costs, we're not just talking about monetary cost, we're also talking about cost in terms of battery usage, display modes, and so on.

Server-side performance

If your app consumes online resources, you'll want to make sure those services are optimized for desktop and mobile users. Fortunately, there's a lot that you can do to optimize performance. In this section, we'll go over at a high level what can be done and where you can go to learn more about enhancing performance on the server side.

Hardware

If your organization has deployed ArcGIS Server on-premises, you have the following options to make your apps perform well:

- Properly sizing your hardware is by far the most important thing you can do to support the number of users your app needs to support. If the app needs to support 1,000 concurrent users but it only supports 50, the user experience will be degraded. To properly size the hardware involves using an Esri tool called the **Capacity Planning Tool** (**CPT**). The CPT is a Microsoft Excel spreadsheet, which is fully explained at the following site: http://www. wiki.gis.com/wiki/index.php/System_Design_Process. With this tool, you can determine the amount of RAM, clock speed, number of servers, and so on by entering a few pieces of information. You can also specify the type of service you plan to deploy along with the factors that affect performance. This tool has a section in it called **Software Technology Performance Factors**, which allows you to test different types of service, as shown here:

Software Technology Performance Factors			
Software	Desktop	Graphics	Density/Portal
AGS103 REST		2D	V ector Only
2.13 Mbpd			
Complexity	%DataCache	Resolution	Output	Client Traffic
Med_Medium	0%	WebDefault	Default	2.13 Mbpd

  As shown here, you can specify the type of server, whether it's 2D or 3D, density, map complexity, percent cache, resolution, and output of services.

- A fast disk subsystem can make all the difference to performance. Work with your IT team or conduct some tests to determine optimal performance if you plan to develop the system from the ground up.

- If your app will publish a geoprocessing service, it's critical that it has the fastest clock speed you can purchase, because these kinds of operation are CPU intensive.

- For memory, Esri recommends 3 GB per core. A four-core server will need 12 GB of RAM.

- Network usage and capacity is critical to the relaying of data between your ArcGIS Server environment and the app. With the CPT, you can evaluate the service times of your app, based on current network use and capacity, and projected network use and capacity.

Data

Geographic data is key to any mapping app so this section will list several techniques for making your app as fast as possible:

- Only add the layers that your app needs. It's tempting to add lots of layers because content is important. However, the more layers you add, the slower the app will be. Esri recommends adding layers in the 1s and 10s of layers.

- Cache your dynamic map services so that users see the data as images at the scale range where the user only needs to see the data.

- Add scale ranges to your layers so that they only show at the scale that they are needed.

- Evaluate the image format of image services, tiles services, and dynamic map services. The CPT allows you to determine the optimal format for these kinds of service.

- With the CPT, you can also test out different resolutions. If your app is going to be mostly used on a phone, choose a resolution of 256x256 or 400x300.

- If possible, generalize layers that have a large number of vertices. For example, if you need to display Chesapeake Bay and your layer has 3 million vertices, consider generalizing it to enough vertices so that it maintains its overall shape, while also removing vertices that aren't needed so that performance is maximized.

- Keep your data in the same spatial reference. Changing spatial references impacts performance, so if the data is all in the same spatial reference, there will be fewer on-the-fly re-projections.

- Optimize the performance of your enterprise geodatabase by following best practices for your DBMS. Consult your DBMS documentation for performance guidelines.

- Use definition queries on your layers when you set up the map to publish.

- Simplify the layer symbology. Avoid using complex symbols because they take longer to draw.

Please consult Esri's help for more details.

ArcGIS Server configuration

When it comes to ArcGIS Server, there are a few key tips to maximize performance:

- It's important that you have enough servers to handle incoming requests. With small setups, a single server will suffice, but with even a few dozen concurrent users, you may be required to have multiple ArcGIS Server installations.

- Increase the number of instances of your services so that more than one instance is handling an incoming request. For more details, navigate to `http://www.wiki.gis.com/wiki/index.php/Server_Software_Performance#Service_instance`.

Client-side performance

Now that we've reviewed some things you can do on the server side, let's turn our attention to what we can do to make ArcGIS Runtime perform optimally. In order to discuss this subject, we're going to have to take a deeper dive into the rendering engine, and that means we will discuss **Graphics Processing Unit (GPU)**, **Central Processing Unit (CPU)**, and the decisions that one must make when deciding on how to render layers. Before we do that, however, we first need to understand a little about graphics programming APIs such as OpenGL and DirectX. ArcGIS Runtime SDK for .NET actually uses DirectX, but we're only going to discuss this at a high level, so these comments generally apply to OpenGL too.

Introducing graphics programming

When you add a layer to the map or scene using symbology individually or via a renderer, the layer is eventually rendered to the GPU. To understand how this works, we're going to talk very briefly about graphics programming. If you were to build your own mapping API, you would not only have to build your own data formats, tools, and so on, you'd also have to build a means to render this content to a GPU so that it renders it quickly.

This process works by taking the original Esri geometries (with symbology applied) you created in geographic space and converting them into screen space as vertices, as shown here in the first image, in the upper-left corner of this screenshot:

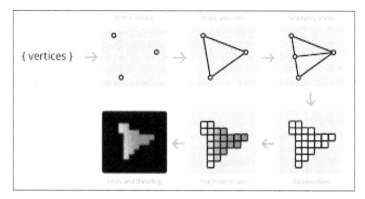

Symbology applied

This presentation is the graphics pipeline. Note that these are vertices for graphics rendering to the GPU, which originally comes from the vertices we created with the Esri geometries. Once the vertices are in screen space, they are converted into triangles. It doesn't matter if the original geometry was an Esri point, line, or polygon, the symbol used to render them on the map is eventually converted into triangles. Here is an example of a polygon that has been converted into triangles:

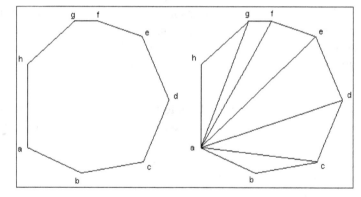

Conversion of polygon to triangles

Once it has been converted into a set of triangles, it is rasterized into pixels, and then the screen coordinates are converted into texture coordinates so that the original Esri symbol is displayed correctly on the GPU. Texturing is the process of converting graphic primitives, such as an Esri `MapPoint` class with a symbol, into a picture that can then be rendered on the GPU. Even if the symbol for the polygon is a simple blue color, this requires a texture that appears blue to be applied to the rasterized triangles (pixels) so that they appear correct to the human eye. Coloring is then applied by the GPU. Once that is complete, the final rasterized image is sent to the frame buffer, where it is rendered to the screen using an OpenGL or DirectX call.

Instead of texturizing the graphics, a more recent advancement has been made, called path rendering. In a basic sense, path rendering is the process of taking text or a graphic and drawing an outline around it, as shown in the preceding screenshot. Path rendering is much more precise-looking than textures, because textures are just pictures while paths are detailed outlines of the text or graphic. See here:

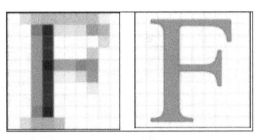

Texture and path

The **F** letter on the left is a texture (picture) while the **F** letter on the right is a path (outline). The outlined **F** letter is much more precise-looking. Path rendering has many advantages and has been created from the preceding graphics.

One other concept that you need to understand is that when you move though the graphics pipeline, this requires changes in state. Every step requires the graphics API to switch modes of operation. Moving from creating triangles to building a texture requires a change in what the GPU is doing. The goal of any graphics engine developer is to reduce the number of state changes that occur because they degrade performance.

Another goal of building a rendering API is to reduce the number of passes that the API has to make when producing an image. Take, for example, a tiled basemap. If you recall from earlier discussions, a tiled service or package is made up of pre-generated tiles that are just images. From the perspective of OpenGL or DirectX, these are just textures, which are added to the display one tile at a time. The primary goal of using OpenGL or DirectX is to reduce the number of times these tiles have to be passed over along with other layers to create a final image. Remember that a nontrivial map will contain several layers that contain basemaps, operational layers, and dynamic layers, which we discussed in *Chapter 1, Introduction to ArcGIS Runtime*. See here:

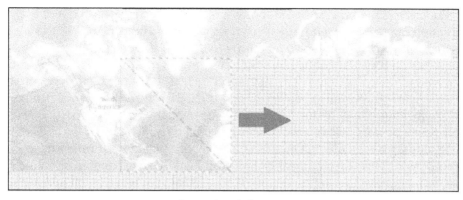

Generation of a basemap

As this basemap is generated, it requires that each tile is created as a texture and placed in its proper location. Each tile is added one at a time until the basemap is finished. Then, depending on how the other layers are configured, the remaining layers may be added while the basemap's tiles are added to the display, or they may not be added until another pass. It can take one or more passes to generate the final image that you actually see on the map. The Runtime Core gives you some control over how many passes it takes to generate an image.

For more information on how the graphics pipeline works with DirectX or OpenGL, see here:

```
https://msdn.microsoft.com/en-us/library/windows/desktop/
ff476882(v=vs.85).aspx
```

Also see here:

```
https://open.gl/drawing
```

Layer rendering options

ArcGIS Runtime comes with two ways of rendering your layers on the map or scene using the CPU or GPU:

- Static
- Dynamic

Static rendering is the process of rendering when the user finishes an operation, such as zooming in. Static rendering is primarily done through the CPU. Dynamic rendering is the process of always rendering to the GPU. Each of these modes has pros and cons to consider. Static rendering does not occur until there is a context change, that is, the completion of an operation such as a pan or zoom. When that occurs, the symbols appear to pop. In other words, they abruptly change from one size to another very quickly during a zoom in operation. The zoom operation is being handled by the CPU, but the rendering is the responsibility of the CPU or GPU, depending on the mode. Once the CPU has completed performing the zoom operation, the CPU takes over in static mode or the GPU takes over in dynamic mode. In other words, the rendering occurs on demand, that is, when a zoom operation completes. When in dynamic mode, this popping sensation does not occur, because the GPU is always rendering the Esri geometry (there is no context change).

Static mode is all about cartographic quality, and that is achieved with path rendering. The entire graphic is rendered as a path. The other big advantage of static mode is that it scales very well. The Runtime Core only regenerates textures when it needs to, when a change occurs in a part of the map. As a result, increasing or decreasing the number of features or graphics doesn't influence performance. The downside of static mode is that it can be memory intensive because it is CPU bound.

For dynamic mode, cartographic quality is sacrificed for improved performance. The reason for this is that textures are used instead of paths. These textures reside in a texture mosaic, which would look something like this:

A texture mosaic is simply a list of textures that come from your symbols in your layers. The texture mosaic always exists on the GPU, which means less work for the CPU, and that means less battery drain. Because it resides on the GPU, textures are very quickly accessed.

To set the rendering mode for your layers, you simply set the rendering mode with `RenderingMode`. Both feature layers and graphics layers support both static and dynamic mode. Tiled layers and map service layers only support static mode rendering, so the dynamic option isn't available for these layer types. When working in dynamic mode, the symbols are converted to graphic primitives, which are then converted into triangles and texturized, as we discussed in the previous section. These symbols are then rendered to the GPU. Static rendering converts the images to paths and renders them to the GPU.

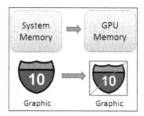

When it comes to text and the labeling of layers, the Runtime Core uses path rendering instead of textures when in static mode. Starting with the 10.2.4 version of ArcGIS Runtime, the Runtime Core now uses texture fonts, which are stored in a texture mosaic on the GPU. Using this approach, the rendering engine can reuse letters on the texture mosaic.

Which rendering mode?

Dynamic mode is best for rendering data during an animation or when navigating. This mode is excellent for tracking yourself using the built-in GPS because the system doesn't have to wait to render. That's because the graphics reside on the GPU at all times in a texture mosaic. As a result, the CPU isn't being used when panning, zooming, or rotating the map. This provides a smooth display experience, but it should be used sparingly when displaying a large number of features. For example, if you have 10,000 points on the map, 10,000 vertices have to be rendered to the GPU, which can result in display degradation. Furthermore, if the symbols are complex, this too can degrade performance. Along with this is that the cartographic quality isn't as good as it is with static mode. But the graphics are screen aligned so that if you rotate the map, the symbols and text stay aligned with how you set them, no matter the orientation of the map.

Static mode, on the other hand, is good for rendering lots of features on the map and achieving high cartographic quality (path rendering). But, because static mode relies more on the CPU, it consumes more of your battery and memory. Furthermore, if one graphic or feature changes, it will then require an update to an entire tile (tiled layer) or map image (dynamic map layer). Furthermore, if the map rotates, the graphics or text also rotate (they are not screen aligned).

So, which rendering mode should you use? Well, it depends on the balance between the number of graphics versus the GPU memory and power. Are you going to be offline? Is battery life a concern? Such similar factors need to be considered. It really comes down to the use case.

Static mode will create the same number of textures no matter the number of graphics, which consumes CPU cycles, which in turn burns battery power. Also, the number of layers and the order of the layers really play a role when in static mode. On the other hand, when in dynamic mode, graphics live in GPU memory all the time, so more graphics means more GPU processing power will be required, and that can result in a slower UI experience. This is especially the case if you have a large number of graphics and a slower GPU, such as on phones. The number of layers and their order is not as important as with static mode.

When in dynamic mode, the number and order of layers isn't that important because resources are shared between all layers on the GPU. It's important that the static layers are placed beside each other on the map, as this reduces the number of passes the rendering engine has to make. If you place a dynamic layer between two static layers, this requires a pass for all three layers. If you place the two static layers beside each other, only two passes are necessary. The reason for this is that the static mode layers don't share resources (textures). Even though the Runtime Core attempts to do some optimizations for you, it's important that you keep this in mind when adding layers to the map.

If a graphics or feature layer contains polygons, static mode is best for feature services if you don't know how many features there are, the kind of symbols used, or the geometry complexity. On the other hand, dynamic mode is good for sketching or interactive layers, real-time feeds, and point layers.

Graphics rendering

Now that we've covered some of the basics of how Runtime Core's display rendering engine works, let's consider how we can optimize graphics layers. Here are some recommendations:

- Use arrays or generics (`List<T>`) when populating a `GraphicsLayer` class.

- Use a renderer whenever possible. The ArcGIS Runtime display engine will look at the graphics in a graphics layer and determine whether they have the same symbol. If they do, the engine will reuse the same texture once, no matter how many graphics are in the layer. If we add graphics individually, this requires converting the graphic to triangles, and then generating a texture. This is a state change. With hundreds or thousands of graphics being added, this also means hundreds or thousands of state changes, which is slower than using a renderer.

Using an ArcGIS Runtime renderer, on the other hand, only requires one state change if there is only one symbol used to make the renderer. As we will see, using a renderer will be much faster than adding graphics individually.

- You can add any kind of geometry to a GraphicsLayer class. However, if you add points and polygons to the same graphics layer, that means that the display engine has to switch states, which reduces performance. It switches states because it has to draw the points, generate textures, draw the polygons, generate the textures, and so on.

- Avoid layer-level transparency because it requires two passes with the GPU, once for the original layer, and then again for the transparency. Symbol-level transparency will be more performant.

Graphics performance testing

Let's now test some different ways of adding graphics to a map so that you can see some differences in performance, depending on how you add the graphics to the GraphicsLayer class. Look at the project called Chapter11c with the code that came with this book. First, let's add 50,000 graphics individually to a map, using a for loop:

```
// get the graphics layer
GraphicsLayer graphicsLayer =
    MyMapView.Map.Layers["graphicsLayer"]
    as GraphicsLayer;
this.random = new Random();

// create a stopwatch
Stopwatch stopWatch = new Stopwatch();
stopWatch.Start();

// create a bunch of graphics
for (int i = 0; i < 50000; i++)
{
    int latitude = random.Next(-90, 90);
    int longitude = random.Next(-180, 180);
    MapPoint mapPoint = new MapPoint(longitude, latitude);

    SimpleMarkerSymbol sms = new SimpleMarkerSymbol();
    sms = new Esri.ArcGISRuntime.Symbology.SimpleMarkerSymbol();
    sms.Color = System.Windows.Media.Colors.Red;
    sms.Style =
        Esri.ArcGISRuntime.Symbology.SimpleMarkerStyle.Circle;
```

```
        sms.Size = 2;
        Graphic graphic = new Graphic(mapPoint, sms);
        graphicsLayer.Graphics.Add(graphic);
}
// stop timing
stopWatch.Stop();
TimeSpan ts = stopWatch.Elapsed;

Console.WriteLine("Time elapsed: {0}", stopWatch.Elapsed);
```

Give this code a go and see how long it takes. Basically, it will add 50,000 points to the map in random locations using latitude/longitude.

Now that we've done this the slow way, let's try a couple of faster approaches. In the next example, let's add 50,000 graphics using a generic list that then has the graphics added to the `GraphicsLayer.GraphicSource` property:

```
// get the graphics layer
GraphicsLayer graphicsLayer =
    MyMapView.Map.Layers["graphicsLayer"]
        as GraphicsLayer;
this.random = new Random();

// create a stopwatch
Stopwatch stopWatch = new Stopwatch();
stopWatch.Start();

// create a enerable list
List<Graphic> graphics = new List<Graphic>(50001);
// create a bunch of graphics
for (int i = 0; i < 50000; i++)
{
    int latitude = random.Next(-90, 90);
    int longitude = random.Next(-180, 180);
    MapPoint mapPoint = new MapPoint(longitude, latitude);

    SimpleMarkerSymbol sms = new SimpleMarkerSymbol();
    sms = new Esri.ArcGISRuntime.Symbology.SimpleMarkerSymbol();
    sms.Color = System.Windows.Media.Colors.Red;
    sms.Style =
        Esri.ArcGISRuntime.Symbology.SimpleMarkerStyle.Circle;
    sms.Size = 2;
    Graphic graphic = new Graphic(mapPoint, sms);

    graphics.Add(graphic);
```

```
}
graphicsLayer.GraphicsSource = graphics;

// stop timing
stopWatch.Stop();
TimeSpan ts = stopWatch.Elapsed;

Console.WriteLine("Time elapsed: {0}", stopWatch.Elapsed);
```

Once you run this example code, you will see that it loads the graphics faster than the previous example, because the graphics are being assigned to the layer as a source directly.

One last example will illustrate adding the graphics using a simple renderer:

```
// get the graphics layer
GraphicsLayer graphicsLayer =
    MyMapView.Map.Layers["graphicsLayer"]
    as GraphicsLayer;
this.random = new Random();

// create a stopwatch
Stopwatch stopWatch = new Stopwatch();
stopWatch.Start();

SimpleMarkerSymbol sms = new SimpleMarkerSymbol();
sms = new Esri.ArcGISRuntime.Symbology.SimpleMarkerSymbol();
sms.Color = System.Windows.Media.Colors.Red;
sms.Style = Esri.ArcGISRuntime.Symbology.SimpleMarkerStyle.Circle;
sms.Size = 2;

SimpleRenderer simpleRenderer = new SimpleRenderer();
simpleRenderer.Symbol = sms;

// create a enerable list
List<Graphic> graphics = new List<Graphic>(50001);
// create a bunch of graphics
for (int i = 0; i < 50000; i++)
{
    int latitude = random.Next(-90, 90);
    int longitude = random.Next(-180, 180);
    MapPoint mapPoint = new MapPoint(longitude, latitude);
    Graphic graphic = new Graphic(mapPoint);
    graphics.Add(graphic);
}
```

```
graphicsLayer.Renderer = simpleRenderer;
graphicsLayer.GraphicsSource = graphics; ;

// stop timing
stopWatch.Stop();
TimeSpan ts = stopWatch.Elapsed;

Console.WriteLine("Time elapsed: {0}", stopWatch.Elapsed);
```

Using the laptop this book was written with, the times for each approach are summarized here:

Added graphics using	Time (in seconds)
Individually (`GraphicsLayer.Graphics.Add`)	0.1866448
`GraphicsSource`	0.1078491
`SimpleRenderer`	0.0429644

These are very fast times, because the laptop used has a high-end graphics card and processor. You may need to reduce the number of graphics when you try it on your PC or laptop. Regardless, from these simple changes we can easily see that when we change from adding the graphics individually to adding them with `GraphicsSource`, the time it takes to render them decreases by about 42 percent. When we go from `GraphicsSource` to using `SimpleRenderer`, the decrease in time is about 60 percent. When we go from adding graphics individually to a `GraphicsLayer` class to using `SimpleRenderer`, the time it takes to render the graphics decreases by about 77 percent. That's outstanding! You will of course get different results, depending on the symbology, number of graphics, and hardware (RAM, processors, and graphics card).

Client-side data access

When it comes to working with layers, there are a few other tips you can use to improve performance:

- Limiting the number of fields in queries. Setting `Query.OutFields.Add` to * is fine for a layer with a small number of fields, but using this approach with a large number of fields will degrade performance.

- Excluding the geometry in your queries can greatly improve performance, especially with several features being returned or with complex geometries.

- Maintaining the same spatial references so that on-the-fly re-projections don't occur.

- When using the routing functionality, you can set the `OutputLines` property on `OnlineRouteTask`. Setting this option can reduce the response size of the data.

- If your app is focused on a particular area of the world, make sure to set `InitialViewpoint` to that area. This can reduce the amount of data coming across the Internet. Note that it's not possible to do this with Windows Store and Phone via XAML; it can be done with code, however.

Summary

In this chapter, we've learned about testing and performance. In particular, we learned about the different kinds of testing with a special focus on unit testing. We first learned about unit testing without MVVM, and then learned how to do unit testing with MVVM Light. We then turned out attention to how to optimize ArcGIS Runtime's performance. In particular, we learned how to optimize ArcGIS Runtime's performance from the server side, and then we turned our attention to client-side performance considerations. Applying this knowledge and these techniques, we can make our apps as fast as possible.

In the next chapter, we'll conclude this book by learning how to build and deploy our app.

12
Configuring, Licensing, and Deploying

Now that we've learned about the API, how to test our app, and how to make sure it is performant, it's time to turn our attention to providing the app to our users. In order to do this, we want to think about options related to configuration, licensing, and deployment. In this chapter, we will discuss these options and provide some guidance on how to best provide your apps to your users so that they can be configured and licensed, and how best to deploy them. We will discuss the following topics:

- Configuration
- Licensing
- Deployment

Configuration

One of the considerations you're going to have to bear in mind is how you want to allow your users to configure the app. In earlier chapters, we used a simple JSON file that was stored locally on disk. While this was fine for learning ArcGIS Runtime, it is by no means a good solution, especially for Enterprise Mobility, where you will deploy your app to dozens or even hundreds of users. How will you address this issue? There are a few options to consider:

- Providing users the ability to configure locally on a device
- Providing users the ability to configure using a web map from Portal for ArcGIS (Portal)
- Providing users the ability to configure on a device and with a web service

Let's discuss each of these so that we understand them. Providing users the ability to configure locally is a nice option, especially if they need to work offline for extended periods of time. Throughout this book, we've shown a simple solution to accomplishing this, using a JSON store. Of course, in order to do this effectively, a UI needs to be built for the app that allows the user to change the app to fit their particular needs and tastes. Of particular importance is how you will provide the user the ability to configure what layers to turn on or off, and how to display them. Do you need to create a symbology tool that allows users to set and save the symbology for each layer or do you just accept the default symbology that was created for the layer in ArcGIS Desktop or Pro?

Another option is to provide users the ability to configure the app using a web map. There are at least two ways of going about this, and each has their pros and cons:

- Using a web map from Portal for ArcGIS (Portal) or AGOL
- Creating a web service that is specific to ArcGIS Runtime

It's very possible to take a web map from Portal and have your app read in its contents to configure your app. This would require that you read in the contents of the web map file stored in Portal's content directory, or by some other automated means, and then load in the configuration to configure the ArcGIS Runtime map or scene with its contents. For example, you could add a layer in Portal, then configure the layer's name, symbology, definition expression, and so on so that your Runtime app could then use those same settings. There are some advantages to this approach:

- You won't have to build your own configuration format or file
- It seemingly aligns well with the notion of the Platform; in this, Portal is the frontend of your Enterprise GIS, while an ArcGIS Runtime app is just one of many consumers of this content
- Most importantly, this approach allows an end user to design the map using Portal for many ArcGIS Runtime users

With that said, there are some cons too:

- A Portal/AGOL web map and the ArcGIS Runtime API don't align their contents perfectly. For example, there's no such thing as `RenderingMode`, `LocalMapServer`, `LocalGeoprocessingService`, and so on in a Portal map. In other words, these are two different concepts that don't correspond perfectly.

- If you change the ArcGIS Runtime configuration locally, how do you apply that change to the web map? Perhaps, you could alter the web map format, but then you risk corrupting the Portal web map. Also, how would you make that change? Obviously, you don't want more than one user making a change to the web map from their device, because everyone would get the same configuration.

- If you upgrade Portal, you'd have to potentially update the web map, which means you'd have to update the way your ArcGIS Runtime app reads in the web map.

A web map is just a JSON file itself, so at first, it seems tempting to use it. For example, here's what the web map looks like:

```
{
    "extent":
    {
        "xmin": -12933906.537,
        "ymin": 3993856.886,
        "xmax": -12933371.998,
        "ymax": 3994375.189,
        "spatialReference":
        {
            "wkid": 102100
        }
    },
    "scale": 1234.5,
    "rotation": -45,
    "spatialReference":
    {
        "wkid": 102100
    },
    "time": [1199145600000, 1230768000000]
}
```

For more information, see here:

http://resources.arcgis.com/en/help/rest/apiref/exportwebmap_spec.html

The other approach is to build a web service, where you build your own configuration file format and expose it as a web service. This is independent to Portal for ArcGIS. This also has some pros and cons. The pros are as follows:

- You can make the file format perfectly match the ArcGIS Runtime API. It would contain configurable items for `RenderingMode`, `LocalMapServer`, and so on that are specific to ArcGIS Runtime.

- With this option, you'd have the ability to allow each user have their own configuration on the server so that they won't have the ability to overwrite a web map as we discussed earlier. In other words, this would be a two-way configuration. A user could update the web service, and then see the change in the app, and likewise, a user could change the app's configuration and see the change in the service.

Some cons are as follows:

- You'd have to build this service and maintain an entire configuration format around the ArcGIS Runtime API. This alone is another piece of technology your team would need to build and maintain.

- As the Esri platform grows and evolves, you may find that Esri eventually develops something like this into Portal for ArcGIS or even ArcGIS Online, so you may find that you're spending a lot of money and time on something that will eventually come as a part of the platform.

In an ideal environment, it would be nice to include the ability to configure the app locally, and then send those configuration changes to the server. Then, if you go offline, you can still make changes locally and keep working. Once you go back online, you can then upload your local changes. Another scenario that may occur is that individual users have more than one device and they want to have two different configurations, depending on what they do with each device. The user could also have more than one ArcGIS Runtime app running on the device too. The user may want application A to have a different configuration on device 1 and 2, and post both of those configurations to a web service. Needless to say, these kinds of options need to be carefully considered, and then implemented so that users have as many options as required, while at the same time having the ability to do their work.

Licensing

One of the great features of developing with ArcGIS Runtime is that you can download it and start developing as soon as you create a developer account. Of course, you will also see this watermark on your map until it is licensed:

You will also see a debug message in the **Output** window of Visual Studio. To have these messages removed from your app, you need to license it so that it meets Esri's license agreement terms. Let's discuss what kind of licensing options we have first, however. You have three licensing options:

- Developer
- Basic
- Standard

You can develop and test using a **Developer** license at no cost. This is the kind of license you have once you download the SDK. A Developer license is not for deployment. A **Basic** license won't cost you anything, except that it has limited functionality. You can deploy with a Basic license. The **Standard** version of ArcGIS Runtime has a cost associated with it and it exposes all of the functionality of the API. You can also deploy with a Standard license. To determine this cost, you need to talk with your Esri representative. Otherwise, you can develop and test using a Developer license, without having to do anything else.

Let's look at the differences between the three license levels:

License level	Functionality
Developer	With this level of license, all functionality is available to you, except that you get the watermark, the debug statement, and the message you've seen with the San Francisco app we've built.
Basic	With this level of license, you get everything except the following functionalities: • Local server • Local routing • Local locators • Local editing • Local geodatabase sync operations
Standard	You get everything.

Refer to the following screenshot for the message that you get in the Developer license:

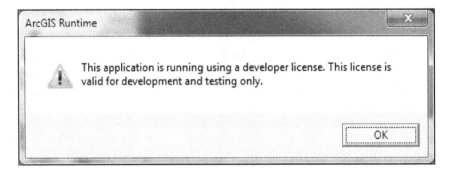

Licensing discussion

So which license do you need? Well, if your app will only consume online services, or just view local content, you can get away with the Basic license. If your app, on the other hand, will do anything with local routing, address geocoding, editing, or syncing with a feature service, you'll need a Standard license.

Licensing the app

As mentioned in the preceding section, you have two options that license the deployment of your app, so in this section, we'll go through the steps for each.

Basic-level licensing steps

To license at the Basic level, you need to go back to the ArcGIS for Developers site:

1. Go to `https://developers.arcgis.com/` and sign in with the account you created at the beginning of this book.

2. Click on **Register New Application**. Complete the fields you see on the form. Then, click on **Register New Application**. You will be taken to the new application's page, where you will see a **Client ID** field, such as this one:

3. Copy the **Client ID** value.

4. In every project file we've created in this book, there has been an `App.xaml.cs` file that contains the start-up code for your app. It looks similar to this:

```
public partial class App : Application
{
    private void Application_Startup(object sender,
        StartupEventArgs e)
    {
        try
        {
            ArcGISRuntimeEnvironment.Initialize();

            //ArcGISRuntimeEnvironment.ClientId = "<Your
                Client ID>";         }
        catch (Exception ex)
        {
```

```
        MessageBox.Show(ex.ToString(), "ArcGIS Runtime
            initialization failed.");

        // Exit application
        this.Shutdown();
    }
  }
}
```

5. Uncomment the line where ClientID is set and past in your client ID from developers.arcgis.com. Consider the following example:

    ```
    ArcGISRuntimeEnvironment.ClientId = "Wa7qYHoBb3FFYOzk";
    ```

6. Run your app and you'll note that **Licensed for Developer Use Only** has been removed. Your app is now licensed for the Basic level licensing.

Standard-level licensing steps

For Standard-level licensing, you have two options:

*   You can use ArcGIS Online or Portal for ArcGIS to license your app. However, this will only work if your app will connect to Portal for AGOL at least every 30 days. This option is a good option for Enterprise Mobility apps.

*   If your app won't be connected at all, or will be disconnected for 30 days or more, you will need to contact Esri for the license code. You can purchase licenses in packs.

Using ArcGIS Online or Portal for Standard-level licensing

To license your app using Portal or ArcGIS Online, you will need to follow these steps:

1. As we did earlier, obtain the client ID for the Basic licensing.

2. You will now need to write some code to have it check ArcGIS Online or Portal. To accomplish this, you'll need to use the IdentityManager class from Esri.ArcGISRuntime.Security. This class allows you to pass in your organizational username/password and it will determine whether your account is authorized. You will also need to add a reference to Esri.ArcGISRuntime.Portal:

    ```
    // Connect to ArcGIS Online or Portal for ArcGIS using
        organization // account
    try
    ```

```
{
    // exception will be thrown here if bad credential ...
    var cred = await
        IdentityManager.Current.GenerateCredentialAsync(
        this.portalURL, this.username, this.password);

    // add the credential if it was generated successfully
    IdentityManager.Current.AddCredential(cred);

    // identity manager will automatically include an
        available credential when connecting (based on the
            server URL)
    portal = await ArcGISPortal.CreateAsync(new
        Uri(this.portalURL));
}
catch (ArcGISWebException webExp)
{
    var msg = "Could not log in. Please check credentials.
        Error
        code: " + webExp.Code;

}

// once connected, get the license info from the portal for
    the current user
var licenseInfo = portal.ArcGISPortalInfo.LicenseInfo;

// set the license using the static
    ArcGISRuntimeEnvironment class
Esri.ArcGISRuntime.ArcGISRuntimeEnvironment.License.
    SetLicense(licenseInfo);

// get a JSON representation of the license info
var licenseJSON = licenseInfo.ToJson();

// ... code here to save JSON license string locally ...
```

This code first uses the `IdentityManager` class to generate a credential based on the URL to a Portal, username, and password. The Portal URL would be as follows if using ArcGIS Online:

```
https://www.arcgis.com/sharing/rest
```

Otherwise, you will need to specify the location of your on-premises Portal for ArcGIS. This will produce a JSON representation of your license.

Now that you have a JSON representation of the license, you need to provide that to your app, as shown here:

```
try
{
    // set the client ID and initialize the ArcGIS Runtime
    Esri.ArcGISRuntime.ArcGISRuntimeEnvironment.ClientId =
        " Wa7qYHoBb3FFYOzk ";
    Esri.ArcGISRuntime.ArcGISRuntimeEnvironment.
        Initialize();
}
catch (Exception ex)
{
    Console.WriteLine("Unable to initialize the ArcGIS
        Runtime with the provided client ID: " +
            ex.Message);
}

try
{
    // use the static FromJson method on LicenseInfo to
        create the
    // license
    var localLicenseInfo =
        LicenseInfo.FromJson(this.licenseJSON);

    // set the license
    Esri.ArcGISRuntime.ArcGISRuntimeEnvironment.License.
        SetLicense(
            localLicenseInfo);
}
catch (Esri.ArcGISRuntime.LicenseException ex)
{
    Console.WriteLine("Unable to license the ArcGIS Runtime
        with the
            license string provided: " + ex.Message);
}
```

With this saved license code, you can now use the app offline for 30 days before you need to reconnect to Portal or ArcGIS Online. If the app is used beyond 30 days, it will only have the Basic license's functionality, which may cause an error if your app has the Standard license's functionality. As such, you will want to not call functionality where the Standard license's functionality is being used. With WPF, you could bind to a property that checks whether the license is valid or not.

To use an extension for a local server, such as 3D Analyst, you will now need to license your app for that extension using the following steps:

1. Add the following code to your app start-up code:

```
try
{
    // set the client ID and initialize the ArcGIS Runtime
    Esri.ArcGISRuntime.ArcGISRuntimeEnvironment.ClientId =
        "Wa7qYHoBb3FFYOzk";
    Esri.ArcGISRuntime.ArcGISRuntimeEnvironment.
        Initialize();
}
catch (Exception ex)
{
    Console.WriteLine("Unable to initialize the ArcGIS
        Runtime with the provided client ID: " +
            ex.Message);
}

try
{
    // use the license code to enable standard level
        functionality
    Esri.ArcGISRuntime.ArcGISRuntimeEnvironment.License.
        SetLicense("runtimestandard,101,rux00000,none,
            XXXXXXX");
}
catch (Esri.ArcGISRuntime.LicenseException ex)
{
    Console.WriteLine("Unable to license the ArcGIS Runtime
        with the license string provided: " + ex.Message);
}
```

2. To specify that the license will use an extension, you need to change the string to something like the following code:

```
Esri.ArcGISRuntime.ArcGISRuntimeEnvironment.License.
    SetLicense(
    "runtimestandard,101,rux00000,none, XXXXXXX",
    "runtimespatial,101,rux00000,none, XXXXXXX",
    "runtimenetwork,101,rux00000,none, XXXXXXX",
    "runtime3d,101,rux00000,none, XXXXXXX");
```

Esri will provide the correct string to replace the x instances. Your app will now run with a Standard license and include the ArcGIS extensions required for the local services.

Attribution

As part of the license agreement with Esri, you need to perform the following steps:

1. Add an **About** screen/dialog and **Credits** screen/dialog, which includes the text Mapping API provided by Esri ArcGIS Runtime SDK for .NET.

2. All map data used in the app must be attributed in the **About** screen/dialog and **Credits** screen/dialog using the **Copyright Text** section of the REST endpoint(s). For example, open the Esri_StreetMap_World_2D map service located at http://server.arcgisonline.com/arcgis/rest/services/ESRI_StreetMap_World_2D/MapServer.

3. You will then need to find **Copyright Text**, which looks like this:

Copyright Text: Sources: Esri, DeLorme, NAVTEQ, USGS, NRCAN, METI, iPC, TomTom

This text must appear in **About** or **Credits** in your app.

4. Finally, you also need to include Esri's logo if your app uses data from ArcGIS Online.

Esri logo guidelines

To meet the requirements for the Esri logo, the app must meet the following guidelines:

- The minimum size of the logo is 43x25 pixels
- The logo must be directly on the map, with no background
- The logo should be located at the lower-left corner of the map or scene
- The logo can't have any transparency applied to it
- Lastly, you can't add another logo, such as your company logo, or any other visual element, on the map or scene, which overlaps Esri's logo

Esri's logos can be downloaded from here:

```
http://links.esri.com/windowsphone/logos
```

Deployment

We have an app now. We're ready to deploy it to the targeted hardware. We've tested it and we've made sure it performs well. The next thing we need to do is figure out how to deploy it to the targeted device(s). Well, one of the great features of ArcGIS Runtime is that it can simply be copied from one device to another and it will just run, without having to create an installation program where you'd have to modify the registry. This of course doesn't include any data your app will need. Also, this doesn't include the fact that developing on Windows Desktop, Windows Store, or Windows Phone requires a different set of concepts and steps that account for the differences in these platforms.

Another great feature of ArcGIS Runtime is that you can pick and choose the options you need to deploy. As a result, if you don't need local services, you don't need to include them. This reduces the size of the deployment, which can be very important with devices such as Raspberry Pi, where disk space will be at a premium.

In the following sections, we'll address the steps you'll need to take to deploy your app.

Deploying with the manifest file

You can create your deployment using the Visual Studio IDE. Let's take the `Chapter1` project that you built in *Chapter 1, Introduction to ArcGIS Runtime*, and modify it so that you can deploy it:

1. Right-click on the project file and click on **Add ArcGIS Runtime Deployment Manifest**, as shown here:

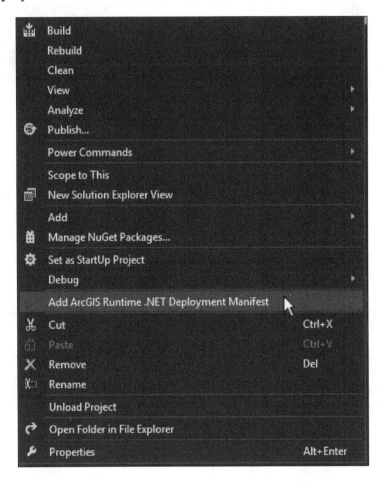

Build	
Rebuild	
Clean	
View	▶
Analyze	▶
Publish...	
Power Commands	▶
Scope to This	
New Solution Explorer View	
Add	▶
Manage NuGet Packages...	
Set as StartUp Project	
Debug	▶
Add ArcGIS Runtime .NET Deployment Manifest	
Cut	Ctrl+X
Paste	Ctrl+V
Remove	Del
Rename	
Unload Project	
Open Folder in File Explorer	
Properties	Alt+Enter

2. Double-click on the resulting file called `Deployment.arcgisruntimemanifest`. It will look like this:

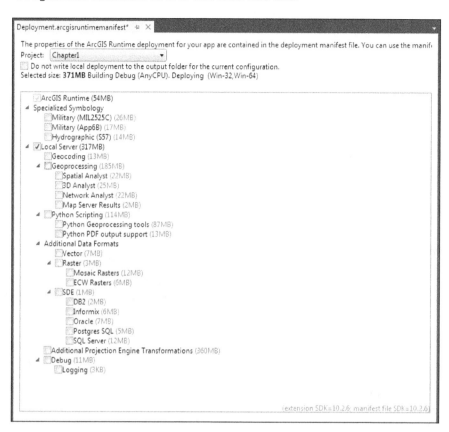

3. You will be prompted to save changes and update your project, as shown here:

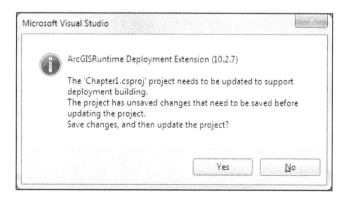

4. You will now be prompted to reload your project, as shown here:

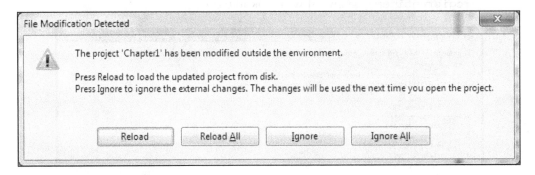

5. Click on **Reload**.

As you can see from this screenshot, you can select the components to include in your deployment. The more options you include, the larger your deployment. A very nice feature with the manifest is that it will total up the size of your deployment for you and it provides the size of each item. If you don't need **Specialized Symbology**, don't include it. The **ArcGIS Runtime** checkbox is also checked; but as you can see in the preceding screenshot, it's only 54 MB. That's incredibly small! You will also note beside **Selected size** that this project is currently in debug mode and it's ready to be deployed as a 32-bit and 64-bit app. If you don't want to include a 32-bit option, change your project's **Build** settings to 64-bit. This will greatly reduce the size of the deployment. If you select **Any CPU**, a deployment for 32-bit, 64-bit, and ARM will be created.

Let's finalize this deployment with a few more steps:

1. Select the option you want to include. If you plan to use local services, you'll need to select **Local Server** and any of the options below it. If you don't select the necessary options for your project, they won't be included.

2. Change your project to run on **Any CPU**, and then build it.

3. Open Windows Explorer and open the directory under C:\ ArcGISRuntimeBook, and you'll note the following directory structure:

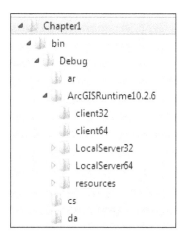

This directory structure shows you that the 32-bit (`client32`) and 64-bit deployments were created. You will also see many other two-letter folder names, such as `cs`, that contain a single file called `Esri.ArcGISRuntime.resources.dll`. These files contain the strings to support ArcGIS Runtime in a specific language.

4. As we are in the debug mode in this app, go to the `Debug` directory and double-click on the app. It's called `Chapter1.exe`. The app will run as it did in Visual Studio.

Deploying without a manifest file

It's also possible to deploy without a manifest file by manually performing the following steps:

1. Create a folder where you want the deployment to be located. Make sure your folder has a descriptive name.

2. Copy the `.exe` file and any required supporting files (the data, 3D models, configuration files, and so on).

3. Copy the ArcGIS Runtime for .NET assembly and the `Esri.ArcGISRuntime.dll` file to the deployment folder. This will be the same location as your `.exe` file.

4. Copy the entire `ArcGISRuntime10.2.7` folder from the `build` directory to the deployment folder.

5. Under the `ArcGISRuntime10.2.7` folder, remove any unnecessary platform folders, such as `Client64`.

6. Verify that your app starts and works properly.

Summary

In this chapter, we discussed some of the configuration options you will need to make for your app. We explored configuring with a local file, a web map (using Portal for ArcGIS or ArcGIS Online) or by creating a custom web service with a format specific to ArcGIS Runtime. We then discussed what you'll need to do to correctly license your app, and the licensing requirements. Finally, we discussed how easy it is to deploy your app when written for Windows.

Congratulations! You've completed this book and you're now ready to explore the rest of what ArcGIS Runtime has to offer, no matter the language you choose, because you now understand the API. With that in mind, check out some more of the sample apps and the other assemblies in the API. You can find lots more non-MVVM samples here:

```
https://github.com/Esri/arcgis-runtime-samples-dotnet
```

At this point, you will have sufficient knowledge to understand the samples and refactor them to the MVVM pattern.

Also, be sure to check out Esri's Developer blog for the latest details about ArcGIS Runtime. The blog can be found here:

```
http://blogs.esri.com/esri/arcgis/category/developer/
```

Index

editing 256
information 256
obtaining 254, 255
updating 258
Augmented Reality (AR) 96

B

barriers 219
basemaps 77
bing layers 81

C

Capacity Planning Tool (CPT)
URL 296
Central Processing Unit (CPU) 298
Chronic Obstructive Pulmonary Disease
(COPD) 261
cities
URL 108
ClassBreaksRenderer 145
client-side performance
about 298
client-side data access 308
graphics performance testing 305-308
graphics programming 298-301
graphics rendering 304, 305
layer rendering, options 301-303
rendering mode 303, 304
compatibility testing 282
Component Object Model (COM) 7
CompositeSymbol 139
configuration 311-314
Controlled Image Base
URL 105
coordinates
converting 131
CreateReplica
URL 232
Create, Retrieve, Update, and Delete
(CRUD) operations 229
CSV layer 81

D

Department of Defense (DoD) 146
deployment
about 323
with manifest file 324-327
without manifest file 327
development testing 282
Digital Terrain Elevation
URL 105
dynamic map service 76

E

early binding 168
editing
about 229, 230
feature service, creating 231
offline editing 231
online editing 231
Runtime geodatabase, downloading 232
envelope 127
Erdas Imagine (IMG) format 105
Esri
URL 1
Esri Developer blog
URL 328
Esri logo
guidelines 323
URL 323
Esri platform 14-17
Esri_StreetMap_World_2D map service
URL 322

F

FeatureLayer class
attachments 253, 254
configuration, editing 241
editing 240
editor 240
edit progress 243
edits, committing 250
features, adding 244, 245
features, deleting 250

output results, working with 275
synchronously 267-269
with ArcGIS Runtime 264-267
online searching
about 181
FindTask 181-184
QueryTask 185-188
task, cancelling 185
Open Geospatial Consortium
(OGC) services 82
OpenStreetMap 83
overlays
about 166, 167
early binding 168
late binding 169-173

P

performance
about 295
client-side performance 298
server-side performance 295
PictureMarkerSymbol 138
PNG
URL 105
polygon 128, 129
polyline 127, 128
Prism 44
projections
URL 68

Q

QueryTask
about 185-188
options 192
using, example for 188-191

R

regression testing 282
renderers
about 141
ClassBreaksRenderer 145
SimpleRenderer 142
TemporalRenderer 146

UniqueValueRenderer 142-145
REST endpoint
URL 271
routes
calculating 222
routing
about 215
app, example 222-228
input parameters, setting up 218
network dataset, setting up 216
overview 217
RPF
URL 105
Runtime geodatabase
creating, from ArcGIS Desktop 236
creating, Geoprocessing tool used 237
downloading 232
downloading, ArcGIS Runtime API used 234-236
downloading, ArcGIS Server REST API used 232, 233

S

scale bar 166
scene example
controlling 109-118
scene layer 82
SDK
and API 17
security testing 282
separation of concerns (SoC) 35
server-side performance
about 295
ArcGIS Server configuration 298
data 297
hardware 296
service instance
URL 298
Shuttle Radar Topography Mission
(SRTM) 105
SimpleMarkerSymbol 137
SimpleRenderer 142
Software Developer Kit (SDK) 1
software performance testing 282

www.ingramcontent.com/pod-product-compliance
Lightning Source LLC
Chambersburg PA
CBHW062053050326
40690CB00016B/3081